peninsula near its ...
centre of population of the district, is situated at the mouth ... the Carrick Roads inlet on the south coast. It was, in the past, the main first-fall port of southern Britain, owing its importance to the protected anchorage of the Roads, and it still maintains a strong connection with shipping through a flourishing ship repair yard, boat-building industry and prospective container-port development. The cathedral city of Truro, situated at the lowest bridging point of the Truro River, developed as the county town because of its road and rail connections with both Falmouth and the Cornish mining district, which was centred on Camborne and Redruth.

The main topographic features of the district, the lower deeply dissected east, dominated by the Fal ria, and the higher open rolling ground to the west, broadly reflect the underlying geology. In the east and north are Devonian rocks — approximately 360 – 380 million years old. They are predominantly clastic sedimentary rocks formed from mud and sand produced by erosion at a time of profound crustal instability and deposited in a deep marine environment. To the west the high ground is underlain by the Carnmenellis Granite and its metamorphic aureole. The granite magma was intruded into the consolidated and deformed Devonian rocks during the Carboniferous Period about 290 million years ago. The metalliferous deposits of the region, copper and tin with subordinate tungsten, lead and silver, form veins (lodes) that owe their origin to the granite. The lodes formed either by deposition from fluids concentrated and expelled during the late stages of consolidation of the magmas, or from fluids rich in metals scavenged from the Devonian rocks during the hydrothermal circulation produced by the heating associated with the intrusion.

The region was subject to erosion and planation late in the Palaeozoic and during the Mesozoic eras but there are no remnants to indicate that Mesozoic rocks may have been deposited over the district. The reshaping of a planated landscape was accomplished in the more recent geological past during the Tertiary and Quaternary periods when there was relative uplift of the land, and retreat of the sea that had covered much of south-west England. In the district there are small areas of unconsolidated Tertiary rocks, remnants that have survived subsequent erosion, stranded on contemporary beach platforms. There are also deposits along the coast and in some valleys that record the active weathering and deposition at the beginning of the last phase of Pleistocene glaciation in Britain. Together with recent sand accumulations and alluvium, such as in the submerged valleys of the Fal ria, these constitute the 'Drift' deposits of the district. Accounts of the consolidated and unconsolidated rocks, their stratigraphy, structure, and metamorphism, the igneous rocks and the mineralisation are included in the memoir. The economic geology of the district is described, and its potential discussed, in the concluding chapter.

Plate 1 Falmouth viewed across Carrick Roads from near Carricknath Point in south Roseland. The Point, in the foreground, comprises interbedded sandstone and slaty mudstone of the Portscatho Formation (A J J Goode).

BRITISH GEOLOGICAL SURVEY

B E LEVERIDGE,
M T HOLDER and
A J J GOODE

Geology of the country around Falmouth

Memoir for 1:50 000 geological sheet 352
(England and Wales)

CONTRIBUTORS

Ore genesis
R C Scrivener

Hydrogeology
R A Monkhouse

LONDON: HMSO 1990

First published 1990

ISBN 0 11 884467 9

Bibliographical reference

LEVERIDGE, B E, HOLDER, M T, and GOODE, A J J. 1990. Geology of the country around Falmouth. *Memoir of the British Geological Survey*, Sheet 352 (England and Wales).

Authors

B E Leveridge, BSc, PhD
M T Holder, BSc, PhD
A J J Goode, BSc
British Geological Survey, Exeter

Contributors

R C Scrivener, BSc, PhD
British Geological Survey, Exeter

R A Monkhouse, BA, MA, MSc
British Geological Survey, Wallingford

Printed in the UK for HMSO

Dd 291134 C10 6/90

CONTENTS

FIGURES

PLATES

TABLES

PREFACE

The Falmouth district (Sheet 352), which encompasses a major part of the Cornish mining district, has a long history of tin and copper exploitation and dependence upon a healthy metal mining industry for its socio-economic stability. The remapping of the district, therefore, funded jointly by BGS and the Department of Trade and Industry, was important in providing an up-to-date base map and a modern geological interpretation of an area already known to be one of the most heavily mineralised in the world.

The country around Falmouth was originally mapped for the Geological Survey by Sir Henry T de la Beche, the Old Series one-inch Sheets 31 and 33 being published in 1839. The first detailed survey on the six-inch scale was by J B Hill and E E L Dixon in 1897–1903 and the one-inch map was published in conjunction with the memoir in 1906. The resurvey on the 1:10 000 scale in 1979–1984 was carried out by Dr M T Holder, Mr A J J Goode, Dr B E Leveridge and Dr R T Taylor under Mr G Bisson and Dr R W Gallois, District Geologists.

The memoir was compiled by Dr B E Leveridge and edited by Dr P M Allen. Dr R C Scrivener contributed notes on fluid inclusions and ore genesis and Mr R A Monkhouse collated the hydrogeological data. Photographs, now registered in the BGS photographic collections at Keyworth, were taken by Mr M Pulsford and Mr H J Evans.

F G Larminie, OBE
Director
British Geological Survey
Keyworth
Nottingham NG12 5GG

5 March 1990

1:10 000-SCALE MAPS AND REPORTS

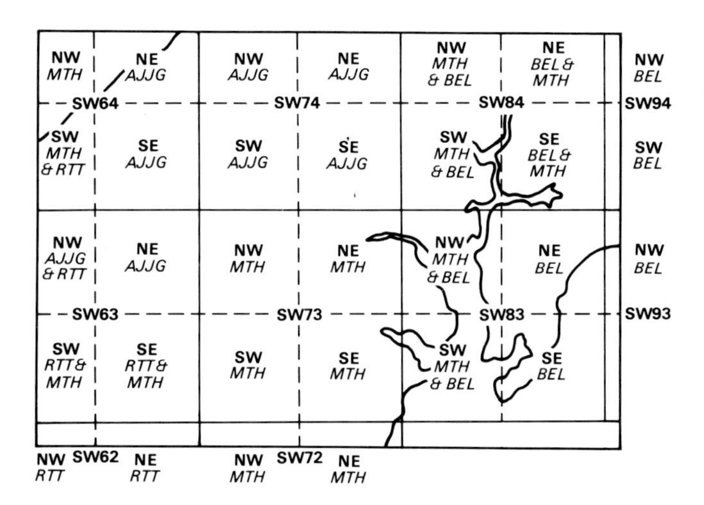

The component 1:10 000-scale National Grid sheets of Geological Sheet 352 are shown on the diagram above with the initials of the surveyors. The surveying officers were M T Holder, A J J Goode, B E Leveridge and R T Taylor. Uncoloured dyeline copies of the maps are available for purchase from the British Geological Survey, Keyworth, Nottingham, NG12 5GG. Areas described in six 1:10 000-sheet reports covering the 1:50 000-scale geological sheet are also shown. Copies of these reports are obtainable from the British Geological Survey, St Just, 30 Pennsylvania Road, Exeter.

NOTES

National Grid references are given in the form [8013 3245] throughout: all lie within the 100 km square SW.

Numbers preceded by the letter E refer to thin-sections in the collection of the British Geological Survey, those preceded by the letters PD refer to internal reports of the Biostratigraphy Research Group of the British Geological Survey.

Dip and strike is recorded and referred to using the 'clockwise' convention where a north-easterly strike signifies a dip towards the south-east and a south-westerly strike signifies a dip to the north-west.

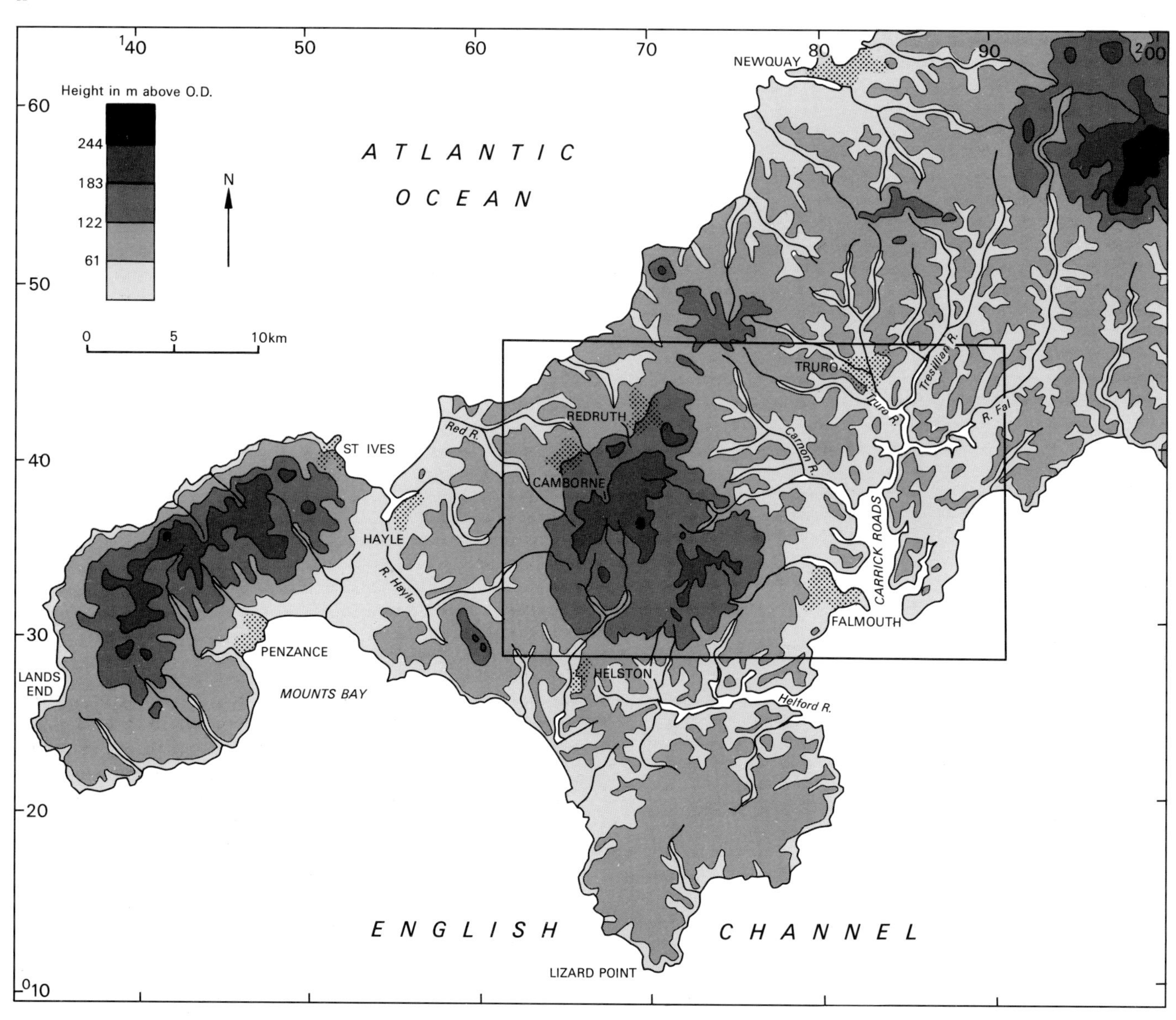

Figure 1 Sketch-map showing the position of the Falmouth district

ONE

Introduction

GEOGRAPHY

This memoir describes the geology of the part of south Cornwall covered by the 1:50 000 scale Falmouth geological sheet (352). The district is one of physical contrasts (Figure 1). Spanning the peninsula, it includes in the north-west an Atlantic coastal section of steep cliffs, 60 – 80 m high. In the south-east, the indented English Channel coast is of gentler aspect; around Gerrans Bay the cliffs are commonly less than 10 m in height. Opening into Falmouth Bay, the Carrick Roads are part of an extensive drowned river valley (ria) system.

The eastern half of the sheet, comprising Devonian sedimentary rocks, is dominated by the ria drainage basin and contains the Fal, Tresillian and Truro rivers. Although lower overall than the ground to the west, the area is deeply dissected. Typically, the steep-sided wooded valleys are separated by gently rounded interfluves. Numerous tidal creeks opening out into the Roads are the drowned lower reaches of extensive dendritic valley systems, which, in their upper parts, are now occupied only by minor streams.

Much of the higher ground in the western half of the sheet is composed of granite. The Carnmenellis pluton forms open country with gently rounded summits rising to a maximum of 252 m above OD. The Carn Marth and Carn Brea satellite intrusions reach 235 and 225 m above OD respectively. Drainage off the Carnmenellis Granite is radial overall, but major joints exert a strong influence on water courses, particularly where there has been active downcutting by streams in the marginal areas of the main intrusion.

The north coastal strip is planated between 80 and 90 m above OD. Although crossed by the Red River and minor streams with watersheds in the higher ground, the amount of low ground is small. This is attributable to strong active erosion (e.g. cliff collapse at Reskajeage Downs) along the coast, where sedimentary rocks directly confront an Atlantic fetch of several thousand miles.

The belt of metalliferous mineralisation crossing the sheet has the highest concentration of mineral lodes in the 'Old World' and has been exploited since the Bronze Age. Tin and copper production was at its height in the middle of the nineteenth century when up to 50 000 people were directly employed. Now (1987), only two mines, Wheal Jane and South Crofty, are in production. The collapse of tin prices in 1986 has meant that the mines still operating in the district, are under threat of closure, with the end of the mining industry, a major employer in south Cornwall, a distinct possibility.

The other traditional mainstay of the regions economy, agriculture, is also currently experiencing stringent times. Soils of the area are thin and clayey over the sediments, and sandy on the granites, which, unusually in the region, are largely enclosed. The land is thus mainly devoted to pasture or feed cereals for dairy farming, and the introduction of EEC milk quotas has hit the industry hard. Although the north coastal region with its restricted sandy beaches and large areas of industrial dereliction has limited tourist potential, the holiday trade on the south coast is buoyant. Former centres of inshore fishing, Falmouth, Portscatho and St Mawes, in common with the smaller coastal hamlets, are now fully committed to the holiday trade. Beaches on the Roseland peninsula coast are sheltered from prevailing winds, and the estuary creeks provide particularly safe moorings.

PREVIOUS GEOLOGICAL WORK

The area was included in the first major regional geological survey in Britain, by De la Beche (1839), who referred its rocks to his 'Grauwacke Group'. The group, of undetermined age, included a variety of rocks, but principally comprised sandstone and slate. The name originated from an area of similar sequences, the Harz mountains, in Germany. Discovery of rocks containing Lower Palaeozoic fossils (Peach, 1841), in association with conglomerates in the adjacent areas of Roseland to the east and in Meneage to the south-west prompted opposing views from the two great protagonists of 19th century geology, Sir Roderick I Murchison and Adam Sedgwick. Murchison (1846) proposed that the conglomerates were basal Devonian rocks lying unconformably on 'Lower Silurian' (Ordovician) rocks that extended through the sheet area, whereas Sedgwick (1852) referred the rocks around Gerrans Bay to 'true Devonian', separated from the older 'Cambrian' (Ordovician) rocks by a major thrust.

Murchison's view was perpetuated when the Geological Survey made the first detailed maps of the region at the turn of the century. Hill and MacAlister (1906) placed the unfossiliferous rocks of the Falmouth one-inch sheet (352) area stratigraphically below the fossiliferous quartzites in Roseland and Meneage, whose age had earlier been confirmed as Ordovician (Collins, 1879). They established a sequence that comprised, from the base, the Lower Palaeozoic Mylor Series (Mylor Slates in Reid and Flett, 1907), Falmouth Series, Portscatho Series and Veryan Series unconformably overlain by the Devonian Grampound and Probus Series. This view, that most of the regional sequence was Lower Palaeozoic, was probably not fully accepted by the Survey because the Mylor Series, Falmouth Series and Portscatho Series were left undated when the quarter-inch map of the district was published in 1909.

In her early papers on the stratigraphy of the region, Hendriks (1931, 1937) amalgamated the Falmouth, Portscatho, lower part of the Veryan and the Grampound and Probus series into a new grouping of Devonian age, the Gramscatho

Beds. This was based on their similar lithologies, said to characterise flysch deposits, and the contained plant remains, which Lang (1929) had identified as being not older than Middle Devonian. Hendriks termed the upper part of the Veryan Series, which included conglomerates and pillow lavas, the Gidley Well Beds, and these were regarded as Upper Devonian rocks that overlapped the Gramscatho Beds. The presence within these beds of Ordovician quartzites, on which the sequence had previously been dated, and Silurian limestones (e.g. Bather, 1907) was attributed to major thrusting during the Upper Devonian.

The Lizard and Meneage one-inch sheet (359) was revised by the Survey (Flett, 1933) and republished the following year showing Hendriks' Gramscatho division. The part of the Veryan Series containing Lower Palaeozoic and metamorphic blocks was termed the 'Crush Zone' and interpreted as a sequence disrupted by late Hercynian movements. Later, Flett (1946) suggested that the overthrust Lizard crystalline rocks acted as a plough breaking up rocks to the north.

The thrust model was further developed by Hendriks (1939, 1949, 1959 and 1966), who envisaged a geanticline developing into a major nappe within which the crystalline rocks of Lizard – Dodman – Start were driven over the Gramscatho flysch. Breccias of Meneage and Roseland formed by shearing of the common limb of the recumbent structure. The implied amount of transport was considerable because rocks of the south Cornish sequence were correlated with European rocks.

The Roseland and Meneage breccia belt was subsequently reinterpreted as a sedimentary breccia (Lambert, 1965; McKeown, 1966; Dearman et al., 1969). The Gramscatho Beds and the breccia belt were considered by Leveridge (1974) to form a typical pre-tectonic flysch/wildflysch association of Devonian age, in which clasts of older sedimentary and metamorphic rocks originated from a rising basement source. Sadler (1973), however, regarded the breccia belt as a condensed marginal Ordovician and Lower Devonian succession emplaced onto the Gramscatho Beds and repeated by thrusting. Subsequently the detailed work of Barnes et al. (1979) and Barnes (1983, 1984) indicated that the Roseland and Meneage breccias are stratigraphically above the Gramscatho Beds and form a continuous sedimentary succession with them.

The resolution of the interrelationship of the main stratigraphic divisions of the region, the Mylor Slates, the Gramscatho Beds and the Roseland and Meneage breccias is dependent on palaeontological information as much as an understanding of the tectonic evolution. Consistent palaeontological evidence of the age of the bedded rocks of south Cornwall was first obtained from limestones of the Middle Gramscatho Beds from Pendower Beach in Gerrans Bay. They yielded conodonts of early Middle Devonian (Eifelian) age (Sadler, 1973; Leveridge, 1974). Less reliable identifications of sparse conodonts from rocks thought to be contemporaneous with matrix in the breccias suggested possible upper Givetian (Leveridge, 1974) and Frasnian (Hendriks et al., 1971) ages. Although it has been generally accepted that the Mylor Slates form the basal division of the regional sequence (e.g. Hill and McAlister, 1906; Wilson and Taylor, 1976), there have been tentative proposals (Stone, 1966; Hendriks, 1971) that they are stratigraphically above the Gramscatho Beds. Turner et al. (1979) obtained Famennian (late Upper Devonian) acritarchs and palynomorphs from the Mylor Slates indicating that, at least in part, they are younger than the Gramscatho Beds. However, Rattey (1980) and Rattey and Sanderson (1982) continued to place the Mylor Slates stratigraphically below the Gramscatho Beds contending (Rattey and Sanderson, 1984) that the palaeontological evidence of the ages of the two divisions was incompatible with the structural evidence of dip and younging in the area. The most recent published palaeontological data (Le Gall et al., 1985) for the lower part of the Gramscatho Beds in Meneage indicate a Frasnian age, which is older than the underlying Mylor Slates.

The Carnmenellis Granite was subdivided petrographically by Hill and MacAlister (1906) and by Ghosh (1934). Subsequent geochemical work by Chayes (1955) and Al Turki and Stone (1978) could not substantiate two of Ghosh's types. The origin of the alkali feldspar megacrysts has been discussed by many authors including Ghosh (1934), Stone (1979) and Dangerfield and Hawkes (1981). Geochemical studies have been carried out by Alderton et al. (1980), Exley et al. (1983) and Charoy (1986). Charoy proposed extensive hydrothermal reworking of the granite creating conditions in which extensive hydrothermal alteration, and the accompanying mineralisation detailed by Dines (1956), took place. Rb/Sr dating of the Carnmenellis Granite has been affected by this alteration such that whole rock ages reported by Charoy (1986) and Darbyshire and Shepherd (1985) are younger than mineral ages. Subsurface tabular shapes of the granites of the south-west England batholith have been suggested by the gravimetric and refraction seismic studies of Bott et al. (1958), Bott and Scott (1964), Bott et al. (1970) and, more recently, Brooks et al. (1984). Both Turner (1968) and Rattey (1980) proposed an intrusion mechanism for the granite that involved an initial updoming of the country rocks over the rising batholith and a subsequent cross-cutting intrusion by the individual plutons. Magnetic anisotropy studies of the granite envelope by Rathore (1980) suggest that some shouldering aside of the country rocks accompanied this later stage of intrusion.

TECTONIC EVOLUTION

The Devonian rocks of the Falmouth district form thick successions, predominantly of sandstone and slate with subordinate olistostromic breccias. These rocks form a south-east-dipping sequence that shows a general increase in the age of stratigraphic units towards the south-east. Sedimentological evidence of way up, however, indicates younging towards the south-east. To explain this apparent contradiction, Leveridge et al. (1984) proposed that the succession is composed of several major thrust nappes verging north-west (Figure 2).

Deformation in the rocks of the Falmouth district is polyphase (Turner, 1968; Rattey, 1979), but the main metamorphic event, dated by Dodson and Rex (1971) as late Devonian, is associated with the first deformation. It is characterised by a south-east-dipping cleavage and tight to isoclinal sheath folds. Sheath folds form in thrust regimes

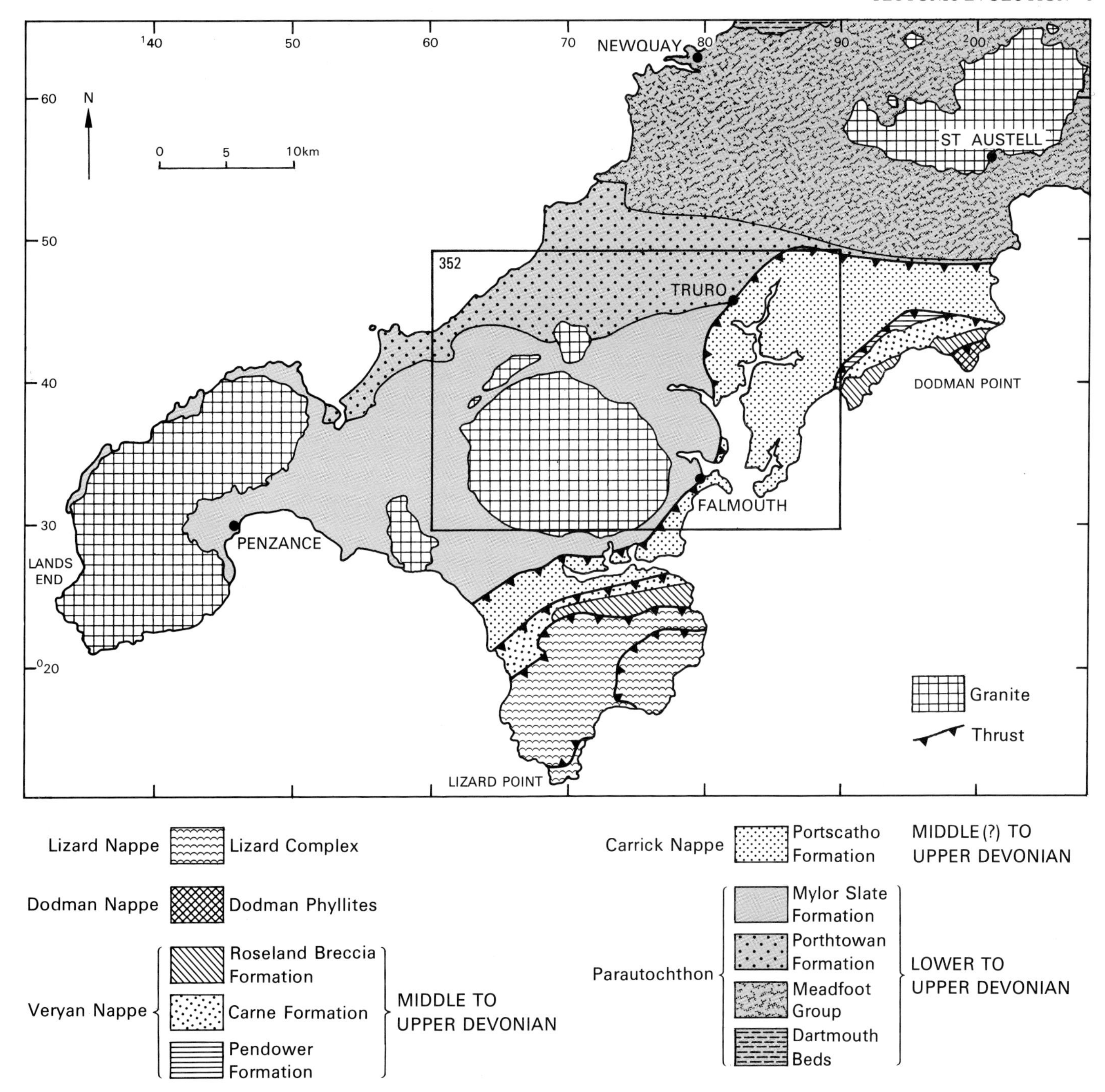

Figure 2 Tectonic units and lithostratigraphical divisions in south Cornwall

(Cobbold and Quinquis, 1980) and here record north-northwest overriding translation (Leveridge, 1974) that may be attributed to the movement of the thrust nappes (Rattey and Sanderson, 1982; Leveridge et al., 1984; Holder and Leveridge, 1986a).

There is evidence offshore of a thrust stack of regional extent from seismic reflection data. Major low-angle reflectors correlate with the Carrick Thrust (Leveridge et al., 1984; Day and Edwards, 1983) and a family of thrusts related to the obduction of the Lizard Complex (SWAT reflection seismic sections, BIRPS and ECORS, 1986).

The presence of pelagic limestone and chert (Pendower Formation) and the ophiolitic remnants (Lizard Complex) with the dominantly flysch-facies sediments in the thrust nappe successions suggests that the turbidite sedimentation took place in an oceanic basin, possibly commencing in the early Devonian and extending into the late Devonian. Modal analyses of the flysch by Floyd and Leveridge (1986) indicate that the sediment originated by the erosion of a continental margin volcanic arc. Sole-mark vectors, the olistolith association in the Roseland Breccias that compares with sequences in Brittany, olistoliths with faunas of

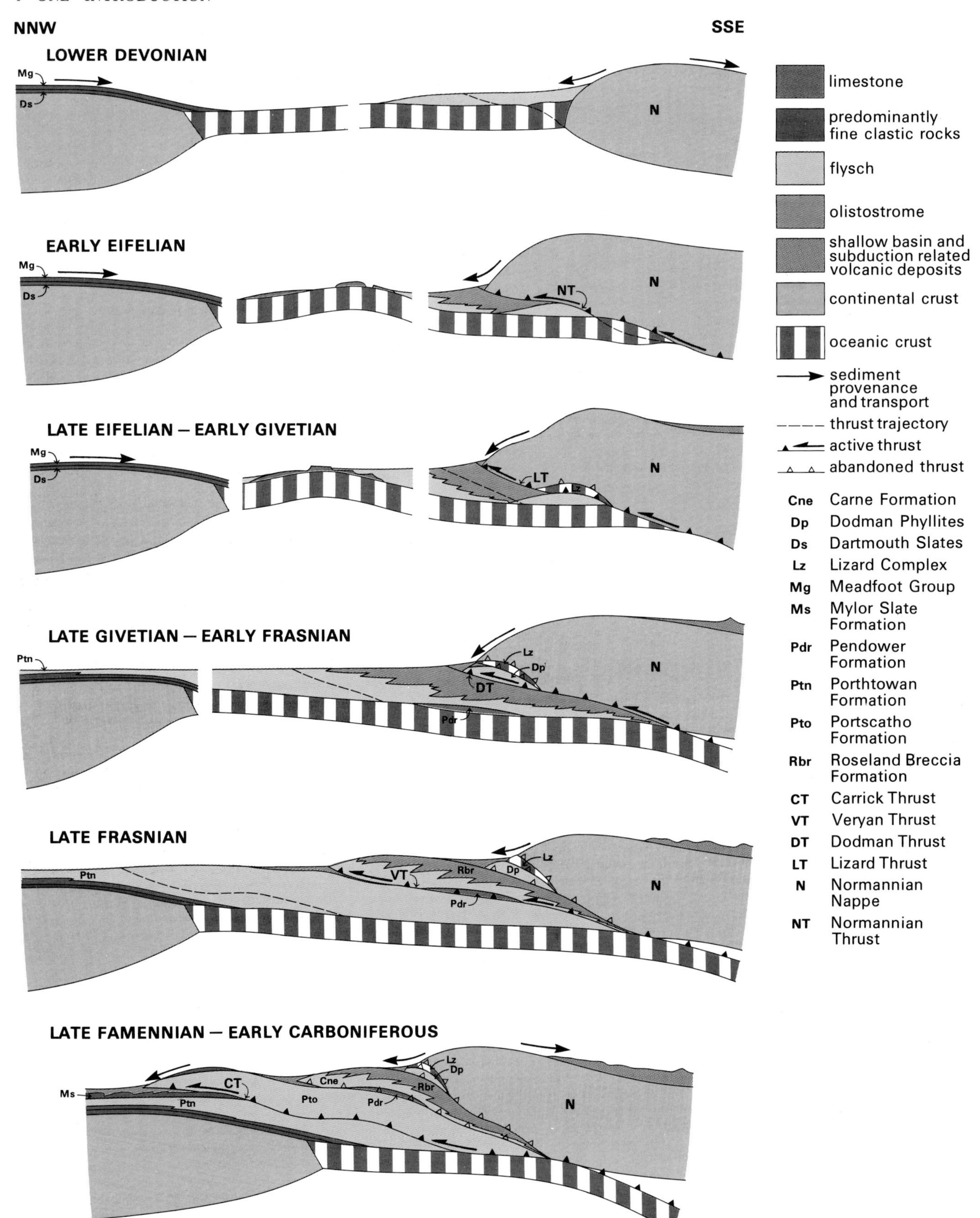

Figure 3 Tectonic evolution of south Cornwall during the Devonian and early Carboniferous

southern affinities (Hendriks, 1937) and the lack of equivalent late Devonian coarse clastic rocks in central Cornwall or south Devon all suggest a southerly provenance for the Devonian sediments. The source area, which also supplied sediment to northern Brittany from Gedinnian times (Renouf, 1974), was named the Normannian High by Ziegler (1982), and said to be the westward extension of the Mid-German Crystalline Rise (Holder and Leveridge, 1986b). It is considered by Holder and Leveridge (1986a) to be a north-north-west-migrating thrust nappe. In south Cornwall climactic wildflysch sedimentation, dynamic metamorphism (Dodson and Rex, 1971) and the first-phase sheath-folding and thrusting all took place in the late Devonian.

The postulated sequence of events (Figure 3) leading to this major deformation began with the southerly deposition of flysch from the Normannian High into the oceanic basin during the Lower Devonian, at the same time as neritic sedimentation was taking place on the northern passive margin of the basin. Northerly directed overthrusting of the Normannian Nappe commenced in early Eifelian times, initiating southward subduction of the oceanic basin floor. The loading of the crust produced by this overthrusting led to the formation of a foredeep at the thrust front with an associated forebulge at its northern margin. This combination trapped clastic sediments being shed into the basin from the Normannian Nappe, allowing a pelagic limestone platform to develop on the forebulge, and hemipelagic sedimetation to take place further to the north. By late Eifelian times continued motion of the Normannian Nappe had detached a fragment of the descending ocean plate, the Lizard Complex, to form a 'horse' that was itself transported northwards on the Lizard Boundary Thrust over the earlier flysch sediments. Final emplacement was not until 370 Ma (Styles and Rundle, 1984). Sedimentation of flysch and olistostromes from the front of the Normannian Nappe and the superincumbent volcanic arc had, by early Givetian times, filled the foredeep and inundated the hemipelagic sediments of the Pendower Formation. During Givetian times, sticking on the Lizard Boundary Thrust, and the higher parts of the Normannian Thrust, caused forward thrust propagation, detaching the early Devonian flysch of the Dodman Nappe and transporting it northwards over the olistostromes of the Roseland Breccia Formation. By late Givetian times flysch and olistostrome sedimentation from the thrust front had filled the oceanic basin, and flysch constituting the Porthtowan Formation had extended onto the northerly-derived neritic sediments of the northern passive margin which was undergoing gradual regional subsidence.

During Frasnian times, forward propagation of the thrusts transferred motion to a new thrust below the hemipelagic sediments of the Pendower Formation, and together with the overlying flysch of the Carne Formation and the Roseland Breccia Formation, this Veryan Nappe was overthrust onto the more distal flysch in the basin, carrying the higher nappes in 'piggy-back' style. Thrust sticking resulting in forward propagation took place again in the late Famennian to early Carboniferous, when the Portscatho Formation flysch of the Carrick Nappe was thrust up out of the basin. Erosion at the thrust front of this nappe produced the thick olistostrome sequence of the Porthleven Breccia Member that was deposited on the distal turbidite sequence of the Mylor Slate Formation. The olistostrome was overridden by continued thrust movements (Leveridge and Holder, 1985), which event marked the final closure of the oceanic basin.

Although the oceanic basin closed at the end of the Devonian Period, and interplate movement was probably largely transferred to oblique dextral strike-slip faulting (Holder and Leveridge, 1986b), tectonism continued to migrate through the peninsula during the Carboniferous period (Dearman et al., 1969; Matthews, 1977). In order to accommodate supracrustal shortening Shackleton et al. (1982) postulated a low-angle sole thrust between basement and Upper Palaeozoic cover. A change from relatively thick thrust nappes from within the flysch basin to thin nappes on the adjoining continental margin is suggested by the change from relatively high-angle ramp reflectors recorded offshore by SWAT (BIRPS and ECORS, 1986) to the shallow subhorizontal onshore reflectors recorded by Brooks et al. (1984). It is most probable, therefore, that during Carboniferous times the southern plate was pushed over the stripped basement of the northern continental margin. The loading produced flexuring and faulting of the underthrust plate (cf. Mitchell, 1974). With relaxation of stress towards the end of the Carboniferous period, movement on the basement faults, and backslip to the south on the major steep thrusts gave rise to a limited extensional regime. Watson et al. (1984) argued that it was under such conditions that the south-west England granites were derived from the lower crust. This may explain their intrusion at the front of the major thrust nappe pile.

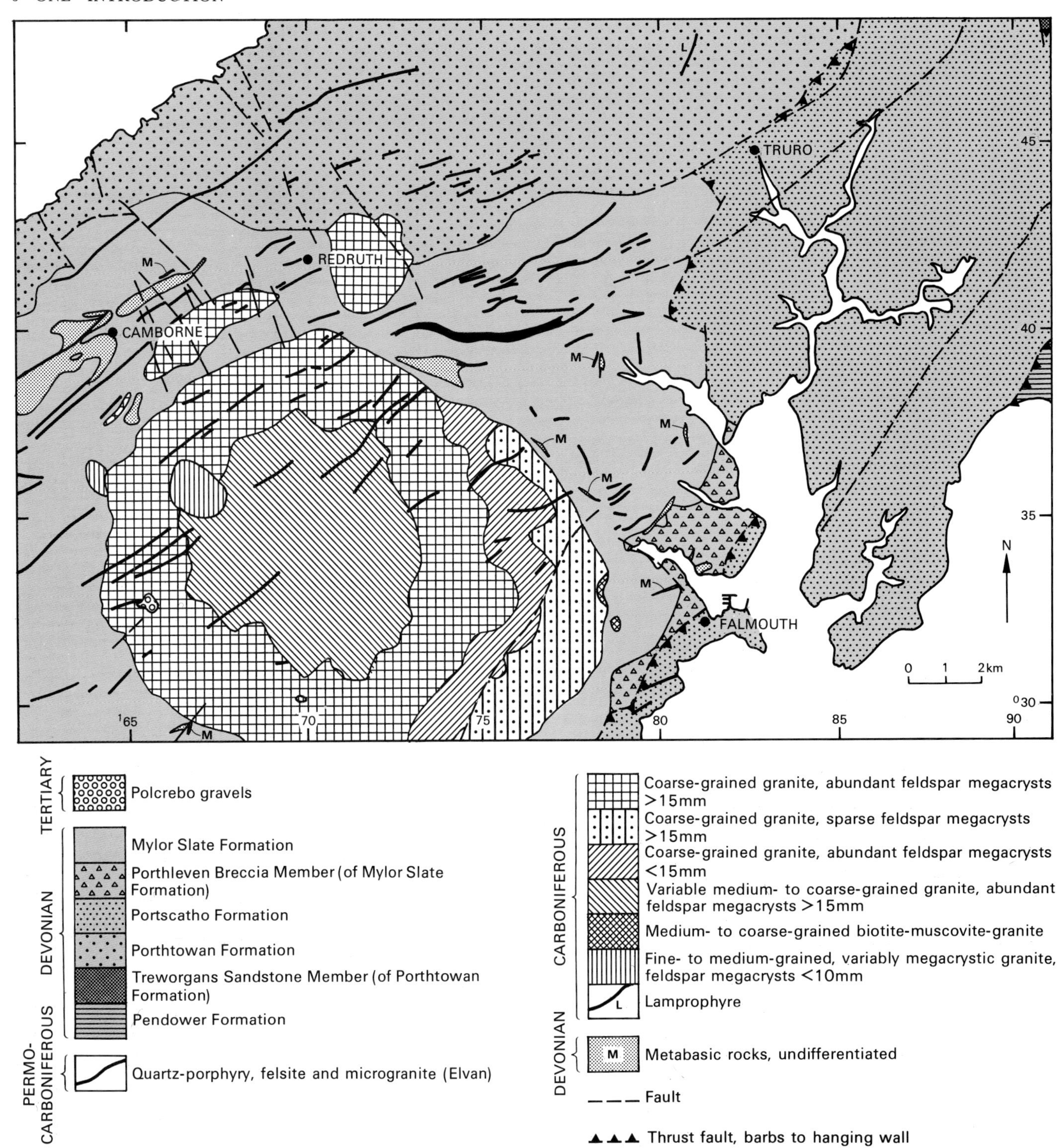

Figure 4 Sketch-map showing the generalised solid geology of the Falmouth district

TWO

Devonian rocks

The relationship between the main Devonian lithostratigraphic divisions of the district, has been reinterpreted by Leveridge and others (1984) and Holder and Leveridge (1986a). They proposed that the Mylor Slates are stratigraphically the uppermost formation in a parautochthonous region that was overridden by the Carrick Nappe, itself part of a thrust-nappe stack containing the obducted Lizard Complex ocean floor ophiolite segment (Strong et al., 1975; Floyd, 1984).

Holder and Leveridge (1986a) included their Gramscatho Group and the Roseland Breccia Formation within the Carrick Nappe. However, new palaeontological evidence obtained by BGS (PD/86/65) supports the conclusions of Le Gall et al. (1985) that the upper part of the Portscatho Formation of that group is of Frasnian age. If the Eifelian conodonts of the superceding Pendower Formation (see Sadler, 1973) are not reworked, then it is probable that there is a significant thrust between the two formations. This thrust, here termed the Veryan Thrust, would separate the Carrick Nappe, newly defined as comprising only the Portscatho Formation, from a Veryan Nappe composed of the Pendower and Carne formations and the superceding Roseland Breccia Formation (Figure 2).

Within the sheet area only the Pendower Formation of the Veryan Nappe, the Portscatho Formation of the Carrick Nappe and the parautochthonous Porthtowan and Mylor Slate formations are exposed (Figure 4).

THE VERYAN NAPPE

Pendower Formation (Pdr)

The Pendower Formation corresponds approximately to the Middle Gramscatho Beds of Hendriks (1937) and includes the Veryan Limestone of Sadler (1973). It has been recognised only in Roseland. The formation is exposed along the Pendower Beach section of the Gerrans Bay coast where it is some 430 m thick. The junction with the underlying Portscatho Formation is seen only at low tide at Pendower [8971 3802] where it appears to be conformable, but new palaeontological evidence suggests that it is a thrust. The junction with the overlying Carne Formation [9087 3809], exposed some 150 m east of Gidleywell, off the sheet, is a clearly defined sedimentary contact.

The formation comprises locally manganiferous mudstone (slate) (Hendriks, 1937) with interbedded and interlaminated sandstone, limestone and chert. The mudstone is variously coloured green, beige, brown, grey or black depending upon association and affinity of composition with the interbeds. Sandstone, limestone and chert form complete and partial cyclic units from a few centimetres to 100 m thick. Each of the lithologies is locally dominant, limestone being so in the upper part of the sequence (the 'Veryan Limestone') where it is common in the smaller rhythms in which sandstone and chert tend to be antipathetic. Chert is prominent in the lowest part of the formation. On the evidence of grading, channelling and other sedimentary structures, the formation as a whole is the right way up.

Conodonts from the limestone (Sadler, 1973; Leveridge, 1974) are mainly of the genus *Polygnathus,* the numbers of which exceed the total of all other genera. These include *Ozarkodina, Bryantodus, Angulodus, Prioniodina, Hindeodella* and *Icriodus*. The age indicated is middle to late Eifelian.

The Pendower Formation contains less sandstone than the other formations of the former Gramscatho Group. The sandstone, grey-green and weathering buff-yellow, is predominantly medium-grained [E 58030] to very coarse-grained lithic greywacke, locally becoming fine breccia [E 59870]. It has a higher content of lithic fragments and lower quartz and matrix proportions than other 'Gramscatho' sandstones (see Figure 5a). Lithic grains are predominantly acid volcanic and schistose metamorphic rocks [E 59869]. The sandstone occurs as graded laminae and beds up to 0.3 m thick, lenses and small channel infills. Fine sandstone laminae and beds of coarse sandstone are interspersed with wispy-bedded sandstone and lenses of very coarse sandstone and fine breccia. The coarse lenses have transgressive bases and in places contain cobbles of limestone and chert. Slump folding and dislocation indicative of soft sediment deformation, are common. Some sandstone is present as isolated blunt-ended phacoids.

Limestone, grey where fresh, and buff or purplish brown on weathered surfaces, is prominent to the east of Carne Beach, but relatively poorly developed between Pendower and Gwendra. The maximum thickness of individual beds is about one metre, although amalgamated beds reach 2 m. At Pendower, some thin beds are discontinuous, comprising isolated lenticles separated by calcareous mudstone. These are unlike tectonically disrupted beds, which have pull-apart veins of quartz and carbonate. Thick beds display bedding characteristics typical of turbidites (Bouma, 1962). The massive, normally graded lower division (A) is succeeded by a division of plane parallel lamination (B), with, in some instances, an upper cross-laminated division (C). Some beds have a traction carpet of fine breccia.

In thin-section the limestone consists mainly of fragments of cloudy micrite, some pelletoidal, crinoid debris, shell fragments, calcispheres and dispersed grains of quartz and quartzite all cemented by sparry calcite [E 59873] (Plate 2a). Grading is determined by the size and concentration of micrite and fossil fragments. The long axes of the fragments are parallel to bedding, and in basal parts of beds coarser fragments are more closely packed than the smaller fragments higher in the beds. Basal fine breccia comprises fragments of limestone with subordinate clasts of schist, phyllite, quartzite, acid volcanic rocks, sandstone and chert with a minor proportion of quartz grains. In general the limestone is little affected by tectonism. The slaty cleavage in adjacent mudstone is represented by widely spaced, irregular

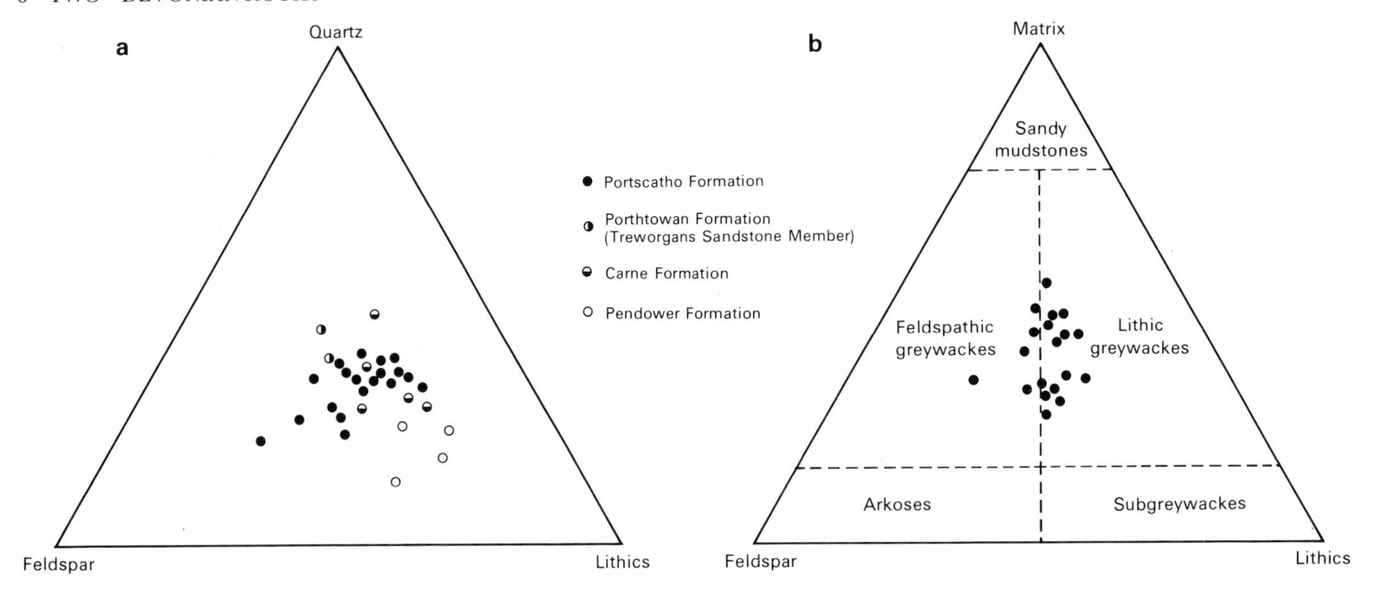

Figure 5 Modal compositions of Falmouth district sandstones (a) and classification of Portscatho Formation greywacke sandstone (b)

pressure solution planes, though in a few cases, shearing is more intense and bands of disrupted calcite crystals enclose augen of unbroken crystals.

Chert units range in thickness up to about 0.3 m but these are made up of several laminae and beds from less than one millimetre to 0.03 m thick [E 59874]. Individual laminae and beds vary in colour from black at the base to grey at the top. This reflects a compositional difference, apparent in thin-section, between finely intergrown microcrystalline quartz, clay minerals and carbonaceous matter of the base and fine, microcrystalline quartz at the top. The carbonaceous material is associated with radiolaria that are more numerous towards bases. Tests are either microcrystalline quartz that is coarser grained than the host rock, with disseminated carbon segregated at the centre, or they are carbonaceous with clear quartz at the centre [E 59959]. These fossils show considerable deformation with directional extension and flattening parallel to cleavage (Plate 2b). The tectonic fabric is marked by the preferred orientation of white mica, quartz and drawn-out carbon fragments. The main development of chert in the formation is at Pendower where it is associated with a bed of deeply weathered lenticles of basic igneous rock [8996 3816]. This is along strike from the Tubbs Mill greenstone, outside the mapped area, which has been shown by Floyd (1984) to have many geochemical characteristics of ocean floor basalt.

a

b

Plate 2 Pendower Formation lithologies

a) Limestone (grainstone) composed of micrite intraclasts, pellets, calcispheres, shell debris and sparse quartz grains, with overprinting secondary sparry calcite ×50 [E59873]. b) Chert with radiolaria flattened and extended in the cleavage ×100 [E59874].

ENVIRONMENT Sandstone composition indicates rapid erosion of a complex, basement source area. Sedimentary structures record their emplacement by turbidity flows and slurries, the concomitant erosion of intrabasinal deposits, and subsequent instability producing slumping. The limestones were also deposited from turbidity currents. In contrast, the varved radiolarian cherts, representing periodic silica precipitation, would appear to be essentially autochthonous. The rhythmic relationships between sandstone, limestone and chert suggests periodic movement affecting the source area, possibly associated with volcanicity.

The autochthonous radiolarian cherts suggest a depositional environment in a deep pelagic basin (see Jenkyns, 1978). The coarse terrigenes were a precursor to the Portscatho flysch but the overall paucity of sandstone suggests that at that time the basin was partly shielded from the source area. Constituents of the limestone turbidites; micrite, crinoid debris, subordinate shell fragments, calcispheres and conodonts, are compatible with a pelagic carbonate platform provenance (Franke and Walliser, 1983) and this could form a rise that would act as a barrier between the basin and the terriginous source area, allowing only intermittent sandstone deposition via breaching channels at times of resurgence of source and supply. The interpretation of the limestones as turbidites derived from a contemporaneous platform implies that the Eifelian age of the contained conodonts is also the age of the basinal Pendower Formation.

THE CARRICK NAPPE

Portscatho Formation (Pto)

The Portscatho Formation is the major formation of the district. It is restricted to the southern side of the peninsula where, in Roseland, it includes the Portscatho and Falmouth series of the previous survey (Hill and MacAlister, 1906) and equates with the Lower Gramscatho Beds of Hendriks (1937). The Carrick Thrust (Leveridge et al., 1984) defines the local 'base' of the formation, and from north to south across the sheet area it cuts up sequence from near the base of the Porthtowen Formation to the top of the Mylor Slates. Though the upper boundary with the Pendower Formation appears conformable at Pendower, where it is observable only at low-tide level, palaeontological evidence implies that it also is a thrust junction.

The formation is very poorly exposed inland, but is almost continuously exposed along the coast and creek sections. It consists principally of alternating grey or greenish grey sandstone flags and grey mudstone (slate) with sporadic thin siltstone beds. They all weather buff in protected coastal sections and inland areas. However, at St Mawes [8405 3270], near Messack [8388 3700] and Turnaware Point [8375 3775] there are minor sequences of purplish red and green interlaminated siltstone and mudstone associated with brown calcareous sandstone (the former Falmouth Series). The formation attains its maximum thickness in Roseland where, uncorrected for thickening by folding or thinning by flattening and normal faulting, it is estimated to be 5.4 km. Sedimentary structures indicate that the sequence is generally the right way up.

The sandstones are thinly to thickly bedded, up to a maximum 2 m, and are largely turbidites (see Walker, 1978). Bedding, where traceable uninterrupted over several tens of metres of foreshore, is plane parallel. Apart from some amalgamated beds, each bed is succeeded by dark grey mudstone varying from a veneer to a thick bed (Plate 3a). Sole markings are not common, but scour, groove and flute moulds (Plate 3b) have been recorded (e.g. [8916 3784]). Internal structures of the sandstone beds are generally referable to the Bouma (1962) model, but complete sequences are rarely present (see Figure 6). Beds may show delayed or multiple grading, a coarse traction carpet or wavy laminated bedding (Dzulynski and Walton, 1965) where wispy siltstone is interspersed with sandstone. Some beds contain pockets and wisps of coarser-grained sand, apparently associated with disturbed cross-laminated sections. High in the sequence there is a significant proportion of beds that contain clasts, mainly of mudstone, dispersed throughout or at particular horizons within a bed, a feature suggesting remobilisation of previously deposited turbidites and emplacement as grain-flow deposits.

The formation is characterised by fining-up, coarsening-up and mudstone subfacies, the last being more common in the lower part of the sequence, where sections up to 15 m thick containing only sporadic thin beds of fine sandstone are present. An overall upward coarsening and thickening is suggested.

Most sandstone is lithic greywacke with the remainder falling just inside the feldspathic greywacke compositional field (Figure 5b). The rock generally displays distribution grading, with coarse to fine sand and silt grains in a finer-grained matrix. Microcrystalline matrix is commonly quartzo-feldspathic, but may be predominantly micaceous [E 59872, E 59877]. There is no evidence that any of the matrix is the product of selective alteration of labile grains, although syntectonic recrystallisation of all finer-grained material and the formation of preferred orientation fabrics has inevitably obscured original textures. The framework grains of the sandstone comprise approximately equal proportions of quartz (Q), feldspar (F) and lithic fragments (L). Among the lithic grains are fine-grained and microcrystalline quartzo-feldspathic acid volcanic rocks, mudstone, siltstone and sandstone, including quartzite, and quartz and mica schists. Plots of framework modes of the Portscatho sandstones (Figure 5a) lie near the centre of the QFL ternary diagram, comparing with other ancient sandstones of 'dissected continental magmatic arc provenance' (Dickinson and Suczek, 1979; Dickinson, 1982).

Plant debris is dispersed through the sequence, but in the uppermost part of the formation at Pendower it forms 'rafts' of coal up to 0.2 m across resting on the tops of some sandstone beds. Lang (1929) considered that much of the fossil wood present could be assigned to the genus *Dadoxylon*, which he suggested indicated a Middle Devonian age for the rocks. Reassessment of the material by Dr G T Creber for BGS (1984) confirmed the *Dadoxylon (Araucarioxylon)* attribution for material from Portscatho facies rocks near Loe Bar in Meneage, but he concluded that the host rocks are no older than lowest Upper Devonian. Le Gall et al., (1985) have published micropalaeontological data indicating a Frasnian age for the upper part of the formation. This age has been confirmed by BGS determinations on miospores and acritarchs

a

b

Plate 3 Sedimentary structures in sandstone turbidites of the Portscatho Formation: a) Grading, the central bed having a coarse traction carpet, the unit above being composite [8920 3781]. b) Flute moulds on the inverted base of a sandstone bed. Direction of viewing S. [8920 3781].

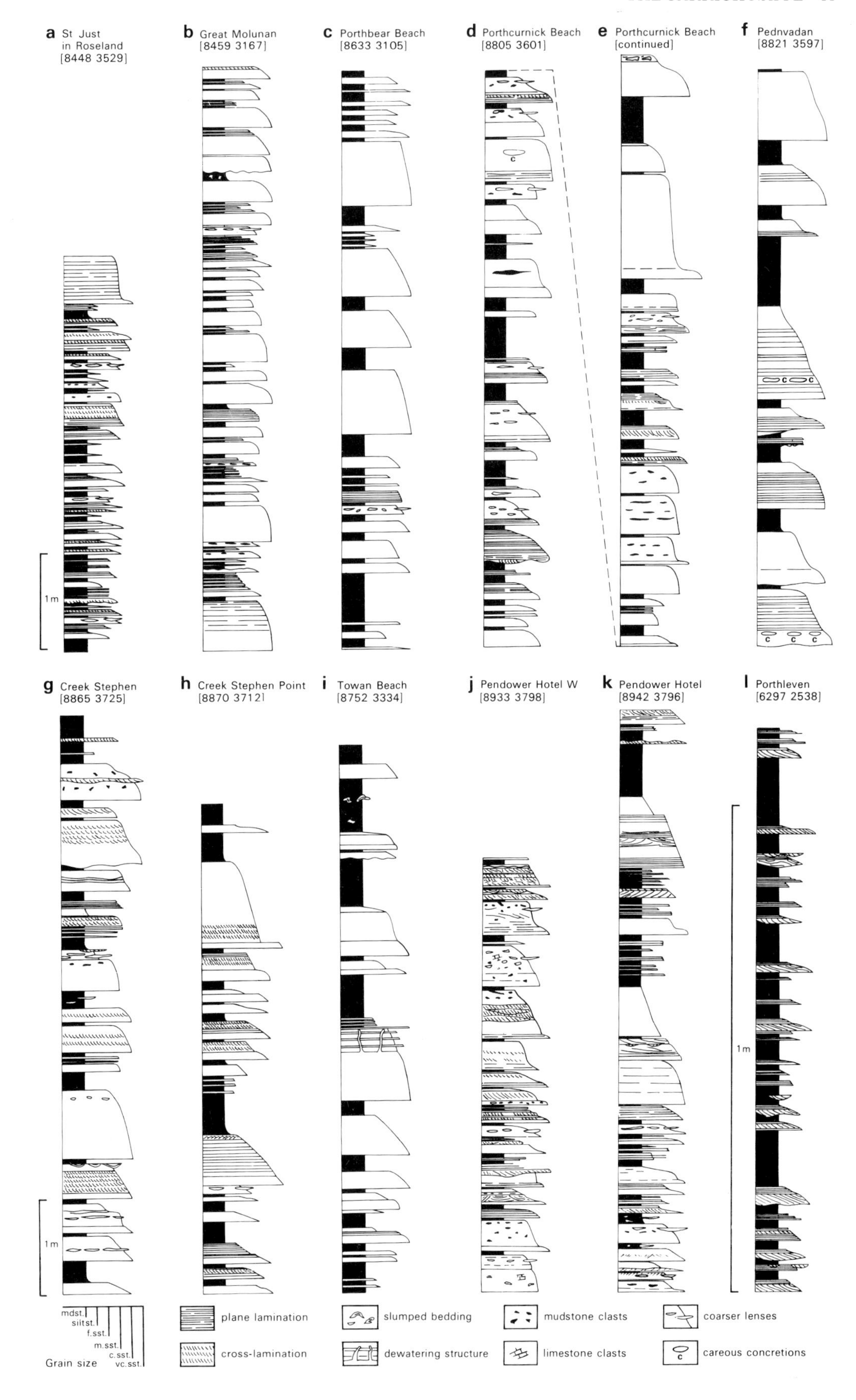

Figure 6 Detailed sedimentary logs of sandstone and slaty mudstone sequences in the Portscatho Formation (a–k) and Mylor Slate Formation (l)

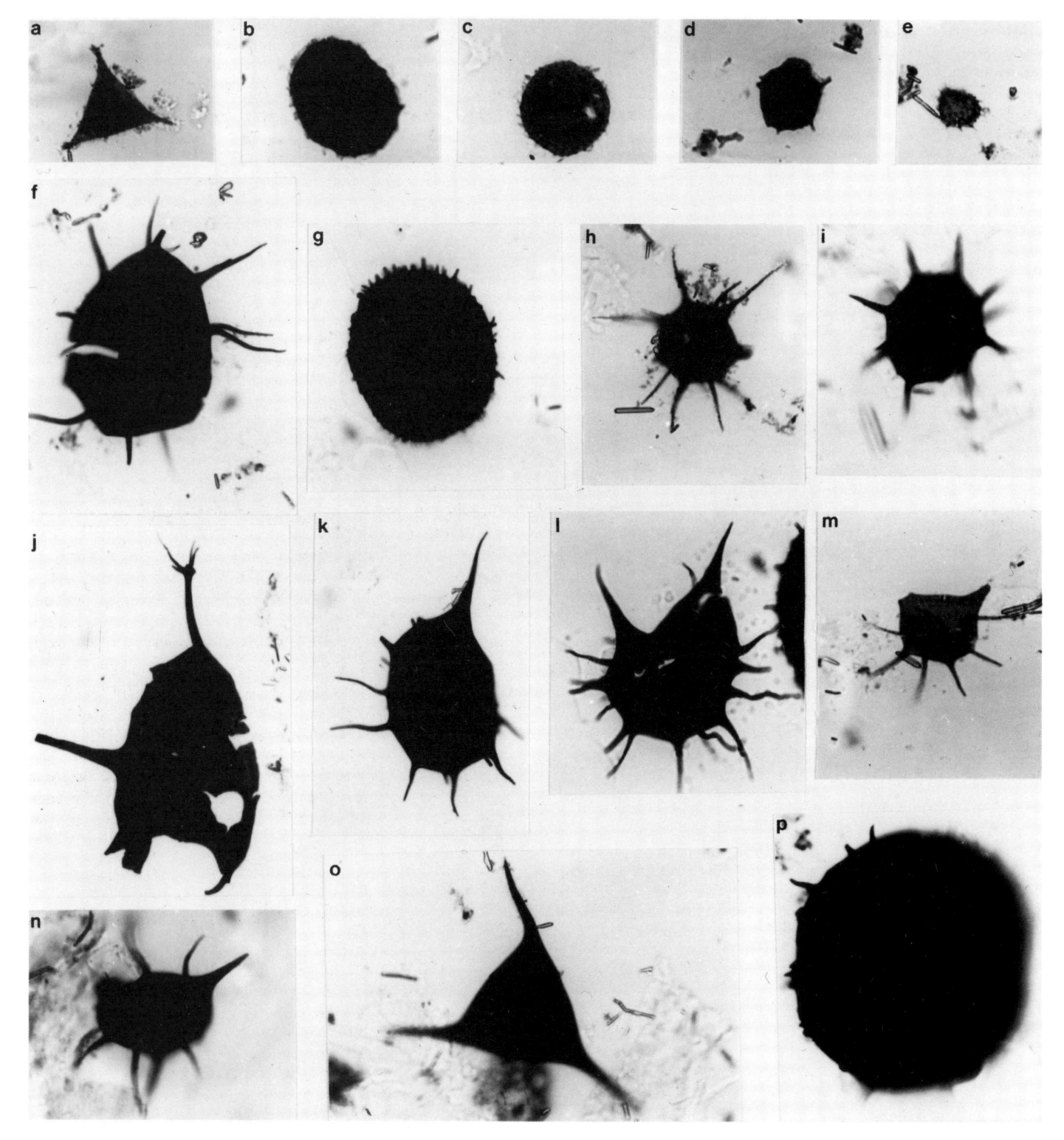

Plate 4 Microfossils from the Falmouth district (see opposite for details)

(PD 86/165) extracted from the mudstone (Plate 4). Most of the mudstone contains carbonised sapropel.

ENVIRONMENT The Portscatho turbidite sequence typifies deposition in a deep-water fan environment (see Rupke, 1977; Walker, 1978), showing a general progradation from outer to mid-fan regimes through the formation. There is further evidence of a deep-water marine character in the dark grey mudstone background sediment containing sapropelic organic residues. The formation is the product of an active tectonic regime recording rapid and persistent uplift of an emergent source area. Concomitant subsidence is necessary to maintain

Plate 4 Microfossils from the Falmouth district
a–i, m–p: Mylor Slate Formation, Mount Wellington Mine
j–l: Portscatho Formation, St. Mawes Castle and Pennance Point

a *Veryhachium ceratioides* Stockmans & Willière. MPK 2559. MPA 5985.
b *Dictyotidium fairfieldense* Playford. MPK 2551. MPA 5984.
c *Micrhystridium dangeardi* Stockmans & Willière. MPK 2552. MPA 5984.
d *Cymatiosphaera polonica* Górka. MPK 2547. MPA 5984.
e *Micrhystridium embergerii* Stockmans & Willière. MPK 2557. MPA 5985.
f *Baltisphaeridium medium* Stockmans & Willière. MPK 2546. MPA 5985.
g *Baltisphaeridium* cf. *echinodermum* Stockmans & Willière. MPK 2550. MPA 5985.
h *Micrhystridium vulgare* Stockmans & Willière. MPK 2555. MPA 5985.
i *Micrhystridium spiniglobosum* Staplin. MPK 2554. MPA 5985.
j *Ammonidium sprucegrovense* ? (Staplin) Lister. MPK 5741. MPA 16117. St. Mawes Castle.
k *Unellium piriforme* Rauscher. MPK 5742. MPA 7131. Pennance Point.
l *Unellium cornutum* ? Wicander & Loeblich. MPK 5743. MPA 7131. Pennance Point.
m *Unellium winslowae* Rauscher. MPK 2545. MPA 5985.
n *Unellium winslowae* Rauscher. MPK 2544. MPA 5985.
o *Veryhachium downiei* Stockmans & Willière. MPK 5744. MPA 5985.
p *Baltisphaeridium simplex* Stockmans & Willière 1962. MPK 2549. MPA 5985.

All specimens ×1200.

Numbers prefixed by MPA refer to residues stored in the National Geosciences Data Centre and those preceded by MPK refer to specimens in the Biostratigraphy Research Group collection, British Geological Survey, Keyworth.

a deep-water depositional basin. Discriminant analysis of mineralogical and geochemical data has led Floyd and Leveridge (1987) to propose the derivation of Portscatho sandstone from a dissected continental margin magmatic arc.

THE PARAUTOCHTHON

Porthtowan Formation (Ptn)

The formation, which includes parts of the Falmouth, Portscatho and Grampound and Probus series of the first survey (Hill and MacAlister, 1906), comprises an estimated 2.8 km-thick sequence of interbedded slate and sandstone. Its base is not on the Falmouth sheet. A transitional junction with the superceding Mylor Slate Formation is recorded in several boreholes in and around the mining district (e.g. Merrose Farm, [6559 4351]). In contrast to the Portscatho Formation, with which it correlates at least in part, grey and grey-green slate is the predominant lithology. In the central part of the formation thinly bedded medium- to fine-grained sandstone and siltstone are present. The beds are graded and many display typical turbidite sedimentary structures, although they commonly lack a massive division at their base. There are thin units of thicker bedded and coarser-grained sandstone throughout the formation, but they are more prominent near the base and at the top. Units showing multiple grading, and individual beds up to 2 m, are present in the upper part of the formation in the cliffs of the north coast (e.g. [6517 4529, 6760 4690]).

The Treworgans Sandstone Member **(TnS)** at the base of the formation is present in the extreme north-east corner of the sheet area, but is not exposed. It is distinctive (Leveridge, 1987), comprising sandstone beds to 6 m, interpreted as grain-flow deposits, interbedded with dark grey slate and wispy interlaminated grey-green siltstone and fine sandstone.

The age of the Porthtowan Formation is not known in detail, but during the survey miospores assigned to the Frasnian (PD 86/135), were obtained from slates at Porthcadjack Cove [6418 4471]. This locality is near the upper junction with the superceding Mylor Slate Formation, which has yielded Famennian palynomorphs (Turner et al., 1979).

ENVIRONMENT The Porthtowan Formation represents a deep-water turbidite fan regime, but the rocks are generally finer grained and appear to be more distal from source than the Portscatho Formation. The coarser-grained turbidite sandstones, however, represent proximal incursions, and a coarsening towards the top may reflect a general progradation. At the base, the finer-grained background sediments of the Treworgans Sandstone Member suggest a shallow, shelf environment similar to the Meadfoot Group (Selwood and Durrance, 1982), whereas the massive sandstones in the member are typical of those associated with deep-water fan facies turbidite sequences (Walker, 1978). Holder and Leveridge (1986a) have suggested that these grain-flow deposits are the products of the breaching and flushing clear of a barrier between the shelf environment and a deeper basin to the south, during northerly progradation of the Portscatho flysch regime.

Mylor Slate Formation (MrSl)

The formation, formerly known as the Mylor Series (Hill and MacAlister, 1906) and Mylor Slates (Reid and Flett, 1907), extends over a large area of south Cornwall. In the type area along the shores of Mylor Creek the formation is not well exposed, but both of the two main sedimentary facies constituting the formation are present. These are interbedded slate and siltstone, and a breccia facies. The formation also includes a substantial igneous component.

The bedded facies of the Mylor Slate Formation is best seen in the Falmouth district in minor foreshore exposures along Restronguet Creek and characterises the Carrick Roads area. Here, dark grey slate with pale siltstone laminae predominates although paler grey slate is present locally [8088 3862]. The laminae are closely spaced, apparently continuous and, where thicker (up to 5 mm), display grading (e.g. [8123 3751]). Elsewhere, thicker laminae and thin beds of siltstone and fine sandstone, showing internal cross and parallel-lamination (see for example Figure 61), typical of distal turbidites, are interbedded with the slate. Siltstone beds up to 0.15 m thick and fine-grained quartzitic sandstone beds up to 0.3 m are present in two areas on opposite sides of the Carnmenellis granite, one south-west of Crowan [647 345], the other north-east and east of Ponsanooth [759 378]. They appear to constitute a separate subfacies within the formation. The Mylor Slate Formation is not closely constrained palaeontologically but assemblages of miospores and acritarchs from Mount Wellington Mine [7758 4128] (Plate 4) have indicated a Famennian age (Turner et al., 1979). These occur within the lower half of the formation, which suggests that it may well extend up into the Lower Carboniferous.

Porthleven Breccia Member (**PBr**)

Hill and MacAlister (1906) interpreted the breccias in the Falmouth district as secondary in origin, but recent work by the Geological Survey has indicated that these and others in the formation are sedimentary (Leveridge and Holder, 1985; Wilson and Taylor, 1976). They are much more extensively developed than suggested hitherto and, although poorly exposed, the presence of beds of breccia through the formation is indicated in spoil and field brash. The breccias near the top of the formation, which form a substantial part of the Mylor Slate Formation in the Falmouth area, have been termed the Porthleven Breccia Member by Holder and Leveridge (1986a).

The member is a complex association of Mylor Slate Formation and Portscatho Formation rocks and a minor proportion of exotic rocks. It has been demonstrated by Leveridge and Holder (1985) that the Mylor Slate Formation lithofacies persists through the Porthleven Breccia member as background sedimentation, whereas the Portscatho Formation lithofacies characterises the macroscopic to microscopic sedimentary clasts. The incorporation of the clasts in the semiconsolidated Mylor Slate Formation locally caused disruption. The resulting melange includes normally bedded Mylor Slates apparently interbedded with Portscatho lithofacies where clast size exceeds that of outcrop. On a smaller scale are framework breccias, matrix-supported breccias and lenses of coarse sandstone. Dark grey laminated slate persists through the member, but it is subordinate to grey-green slate, which in places is interlaminated or interbedded with siltstone or fine sandstone, but generally occurs as matrix to the breccias.

Exposures of the Porthleven Breccia Member in the sheet area are generally poor, but Leveridge and Holder (1985) have interpreted the foreshore sections of interbedded, graded greywacke sandstone and grey slate between runs of breccia at Flushing [818 336] and Falmouth [807 331] and in Falmouth quarry [8048 32991] as breccia containing Portscatho facies clasts several metres in diameter.

Clasts in the mesoscopic range are present in the matrix-supported breccias of the area (Plate 5a). On the western side of Restronguet Point [819 376], blocks of sandstone up to 2 m in length, and smaller blocks of siltstone, mudstone and some trachyte, are dispersed in grey-green slate.

Secondary fabrics are present, but do not obscure the relationship of early (S_1) cleavage to the clasts of the breccias. S_1 tends to penetrate the less competent clasts but pass around more indurated sandstone and siliceous siltstone fragments. Except for some of the larger sandstone phacoids and some pre-cleavage contorted sandstone fragments, the clasts are lenticular in sections perpendicular to cleavage. They show no preferred dimensional orientation, even near the Carrick Thrust where siltstone clasts show a high degree of flattening (Plate 5b).

Igneous rocks in the Mylor Slate Formation

A broad belt of gently inclined greenstones, reported to include basic lavas and volcanic breccias associated with high-level basic intrusions (Floyd and Al-Samman, 1980) are interdigitated with the sedimentary rocks near the base of the Mylor Slate Formation. Overgrown exposure and recent tipping has obscured the evidence of the massive and pillow lavas in the viaduct area [630 391] recorded by Al-Samman (1980) and in a railway cutting south of Gwinear Road [6191 3797] recorded by E E L Dixon during the first survey (Note book, BGS archives). The few available exposures of greenstone range from very fine-grained basalt to gabbro and among field brash are blocks of fine-grained amygdaloidal greenstone. However, none of the rocks has been established as extrusive during the survey and they are described in Chapter 3.

Environment

The Mylor Slate Formation was interpreted by Wilson and Taylor (1976) as a quiet-water marine basinal sequence with distal turbidite incursions. The breccias, which are interbedded with the laminated slate of the the Porthleven Breccia Member, represent a more energetic regime. They are essentially mass flows (see Middleton and Hampton, 1976). The interbedded coarse sandstone lenses probably represent less turbid slurries. Mudstone and siltstone clasts record penecontemporaneous disruption and incorporation of semiconsolidated intrabasinal deposits, a process probably induced by turbulence associated with the introduction by sliding of extrabasinal blocks of Portscatho facies sandstone and slate. The intermittent supply of breccia clasts through several hundred metres of the Porthleven Breccia Member requires an active tectonic escarpment at source (Cowan and Page, 1975). Leveridge and Holder (1985) have proposed that the breccias at the top of the Mylor Slate Formation are derived from the Portscatho flysch in the Carrick Nappe as it was thrust across the Mylor depositional basin at the end of the Devonian.

Plate 5 Porthleven Breccia Member at Flushing

a) Irregular flattened clasts of pale sandstone and siltstone supported in a green-grey slaty mudstone matrix [8074 3395]. b) Strong flattening and disruption of siltstone clasts near the Carrick Thrust [8165 3359]. Polished surface 7 × 5 cm.

THREE

Igneous rocks

Intrusive and extrusive basaltic rocks, lamprophyre, granite and granitic porphyry are present in the Falmouth district. The basaltic rocks are contemporaneous with the Mylor Slate Formation. Lamprophyre dykes and sills are uncommon and occur sporadically within a poorly defined north – south zone extending through Carrick Roads and Truro, although a single dyke is also present at Chyweeda in the west of the Falmouth district. The Carnmenellis Granite and its largest satellites, Carn Marth and Carn Brea, comprise approximately one third of the all the rocks cropping out in the Falmouth area and form part of the major batholithic chain of south-west England. The granite porphyries, locally termed elvans, form part of a province-wide dyke swarm that is particularly strongly developed in the base-metal mineral zone of south Cornwall, crossing the Falmouth district.

BASALTIC ROCKS

Basaltic rocks in south Cornwall crop out in four north-east – south-west-trending linear belts, of which two cross the district. The most northerly of these two extends between Redruth and Penzance, within the outcrop of the Mylor Slates, and comprises pillow lavas, volcanic breccias and fine- to coarse-grained intrusions.

A second belt of basaltic rocks within the Mylor Slates, extends from Restronguet near Falmouth south-westwards out of the area to Porthleven. At the north-eastern end of this belt doming effects associated with the intrusion of the Carnmenellis Granite have swung the outcrop north – south and north-west – south-east. The rocks of this belt, much weathered and poorly exposed inland, are dominantly fine- to coarse-grained intrusions.

All the basaltic rocks are cleaved to some extent and predate the first deformation and regional metamorphism (Floyd and Al-Sammam, 1980; Barnes and Andrews, 1981) dated by Dodson and Rex (1971) at between 365 and 345 Ma. Contemporaneity between these rocks and the Mylor Slates is indicated by vesicular and slaggy margins, autobrecciation and crude pillow structures around some gabbroic intrusions (Taylor and Wilson, 1975). Those intruded into the Porthleven Breccia Member were immediately deformed at the migrating tectonic front.

Despite the pervasive nature of the low greenschist facies regional metamorphism, which has converted the basaltic rocks to greenstones, parts of the primary mineral assemblage in the basic rocks have been preserved (Floyd and Lees, 1973; Floyd, 1983) and sufficient evidence is available to suggest that the primary mineral assemblage was plagioclase, clinopyroxene, ilmenite ± olivine.

Geochemically all of the south Cornish basaltic rocks are tholeiites (Floyd and Winchester, 1975). Although Zr/Y ratios indicate that they span the ocean-floor/intraplate fields, Floyd (1981, 1984) has shown that they can be divided into two geographically separate and chemically distinct groups. The group that is typical of intraplate tholeiites (Al-Samman, 1980) lies within the Mylor Slate Formation. According to Floyd and Al-Samman (1980) in the Falmouth area they are characterised by high Y values. The lavas have been fractionated from the intrusions and are typical of stable continental margin environments. This is in marked contrast to basic rocks within the Veryan Nappe which either have ocean floor basalt chemistry (Floyd, 1984) or are related to the subduction of oceanic crust (Floyd and Leveridge, 1987).

LAMPROPHYRES

The lamprophyres generally are steep, sheet intrusions parallel to the locally dominant cleavage or bedding, but there are also some less steeply inclined, sill-like bodies (e.g. Messack [8395 3660]). Abrupt changes in dip and trend of the intrusions are common. *En-échelon* arrays of dykes are also present. The intrusions are, in general, between 1 and 15 m thick, but the thickness of an individual dyke commonly displays considerable variation associated with changes in its trend. Marginal brecciation of the country rocks occurs around the Pendennis intrusion [827 316] and intrusive breccia dykes are present within it.

The lamprophyres are reddish brown, fine- to medium-grained porphyritic rocks classed as phlogopite and olivine-phlogopite-minettes by Dangerfield et al., (in press). Zoned micas up to 7 mm in diameter, with a phlogopite core and biotite rim, form the main phenocryst phase, although up to one third of the dykes also have olivine phenocrysts psuedomorphed by quartz and iron oxides [E 59870 from the Messack sill]. The groundmass is composed of feldspar laths between 0.2 and 0.5 mm in length, equant biotite flakes and interstitial quartz. The feldspars are generally orthoclase with subsidiary albite, except in strongly weathered lamprophyres where albite is the only feldspar. Although Hall (1982) reported that sanidine was the sole feldspar in the Pendennis intrusion, Dangerfield et al. (in press) have only recognised orthoclase and albite there. Riebeckite and potassio-magnesio-arfvedsonite (Hall, 1982) form 1 per cent of the Pendennis dyke. Small scattered needles or large euhedra of apatite, and opaque iron oxides form the main accessory mineral phases. Monazite, zircon and sphene may also be present and baryte, chromite, hedenbergite, copper sulphides and primary carbonates have been described by Hall (1982) from the Pendennis intrusion.

The lamprophyres are extensively altered. Ferromagnesian minerals have been partially or completely replaced by chlorite and carbonates and, in heavily weathered rocks, by iron oxides and clay minerals. In general, feldspars have been sericitised, chloritised or replaced by quartz and carbonates.

Xenoliths are common and consist largely of locally derived slate, sandstone and vein quartz together with a suite of reddened, coarse-grained, nonmegacrystic granitic rocks and pegmatites (Hill and MacAlister, 1906; Smith, 1929). In the Pendennis dyke Smith (1929) reported '...brown glass... containing crystals of sillimanite, spinel and probably andalusite...'. Sphene and a green pyroxene within the margins of a vein quartz xenolith are also reported by him from the Trelissick dyke [835 391] where the aluminosilicates appear to have been formed by metamorphism and/or metasomatism after inclusion within the lamprophyric magma.

Geochemical data presented by Hall (1982) for the Pendennis lamprophyre indicate that it is oversaturated and peralkaline, despite the inclusion of Al-rich sedimentary rock as xenoliths. There is also enrichment in 'mantle incompatible' elements (P, Ba, Sr and REE), similar to the enrichment patterns of kimberlites and carbonatites.

The field relationships of the lamprophyres provide few constraints on their age. They all appear to be post-tectonic, but their relationship with the granite is unknown. Collins and Collins (1884) reported that a dyke at Treliske [799 450] was intersected by a quartz-porphyry (elvan) dyke, an observation not subsequently confirmed. The radiometric ages of the lamprophyres (296 ± 2.5 Ma by K-Ar method, Dangerfield and others, in press) and elvans (280 – 270 Ma by Rb-Sr method, Darbyshire and Shepherd, 1985) provide some supporting evidence. The intrusion of the lamprophyres is thus broadly contemporaneous with that of the Carnmenellis Granite.

GRANITE

The Carmenellis Granite is broadly circular in outcrop and is surrounded by a series of satellite granite bosses that vary in size from a few tens of metres to several kilometres in diameter. Although in general the granite is a coarse porphyritic muscovite-biotite-granite, there are internal variations in texture. Hill and MacAlister (1906) described the Carnmenellis pluton as a porphyritic orthoclase-perthite, muscovite-biotite-granite that was 'fairly uniform throughout', but drew attention to the finer texture of material in the eastern zone and to two patches of fine-grained granite that crop out near Bolitho and Praze-an-Beeble.

Ghosh (1934) agreed with the general disposition of the fine-grained granite mapped by the Survey, but divided the main porphyritic mass into three components. Type 1, a coarse megacrystic granite, equated with the Survey's normal porphyritic variety, forms an annular outcrop in contact with the country rocks over much of the granite contact. Type 2, regarded as intrusive into Type 1 and of generally finer texture, forms an arcuate outcrop occupying most of the eastern part of the granite mass. Type 3, a sparsely porphyritic fine-grained granite also thought to be intrusive into Type 1, forms a circular outcrop in the centre of the pluton. Ghosh (1934) also grouped aplitic veins and sills, and the quartz-porphyry (elvan) dykes together as a late-stage phase of minor intrusive activity, but he considered the fine-grained granite near Bolitho and Praze to be connected to the Type 1 granite. Chayes (1955), in comparing modal analyses of the granite types mapped by Ghosh (1934), found no difference between the Type 1 and 2 granites. Stone and Austin (1961) suggested that all three granite phases were local variations of a single intrusive unit. However, Al-Turki and Stone (1978), using a statistical analysis of modes and chemical analyses, confirmed Chayes' (1955) result, indicating that the Type 1 and 2 granites displayed the same modes and chemistry, with the sole exception of their Rb contents. Type 3 showed differences to 1 and 2 in all respects.

The present survey has divided the Carnmenellis Granite into seven mappable varieties (G_a to G_h, Figure 7a) distinguished from each other on textural grounds. The outcrop pattern of the varieties is in broad agreement with that of Ghosh (1934).

G_a This is the predominant textural variety (Plate 6a, 6f). It corresponds broadly with Ghosh's (1934) Type 1 granite, forming a nearly circular, annular outcrop in contact with country rock on the north, west and south sides of the pluton (Figure 7a). The large satellite intrusions, Carn Marth and Carn Brea (Plate 6f) are also of this variety. Some diffuse patches of G_a occur in the G_c granite at Carnsew Quarry [762 346].

The G_a granite is composed of 15 – 20 mm alkali feldspar megacrysts in a coarse-grained matrix of alkali feldspar, plagioclase, biotite, quartz and white mica ± chlorite, epidote and tourmaline [E59907, E59908]. Alkali feldspar forms abundant megacrysts up to 20 mm long, which in some places show a pronounced linear or planar alignment, and 2 – 4 mm ragged grains in the groundmass. Both forms are perthitic and partially altered to white mica. The megacrysts are twinned and contain inclusions of 0.5 – 1 mm euhedral plagioclase, rare biotite or chlorite/white mica aggregates and quartz, present as rounded inclusions at the margins of the megacrysts or vermicular intergrowths within them. Alkali feldspar in the groundmass is associated with and partially encloses plagioclase and biotite, and is intergrown with angular quartz grains in the interstices of plagioclase/biotite aggregates. Alkali feldspar is strongly replaced by vermicular albite at plagioclase/alkali feldspar contacts and in some places by round quartz grains. Elsewhere, white mica and subhedral, zoned, yellow, pleochroic tourmaline replace it in the groundmass.

In general quartz has no inclusions and is interstitial to biotite and the feldspars, occupying large pools up to 6 mm in diameter between biotite/feldspar aggregates.

Somewhat sericitised 2 mm euhedral laths or 4 mm anhedral grains of plagioclase with multiple twinning and localised compositional zoning are also present in the groundmass. Both anhedral grains and laths occur as aggregates partially enclosing biotite or its replacement minerals.

Biotite is brown, strongly pleochroic with abundant pleochroic haloes and 0.4 mm apatite inclusions. It is generally present in clusters of flakes 3 – 5 mm in length either slightly chloritised or completely replaced by a felted mat of chlorite and opaque minerals, epidote and white mica.

Dark xenoliths present in the G_a phase [E59906], are limited to a few 30 – 50 mm diameter subangular xenoliths at

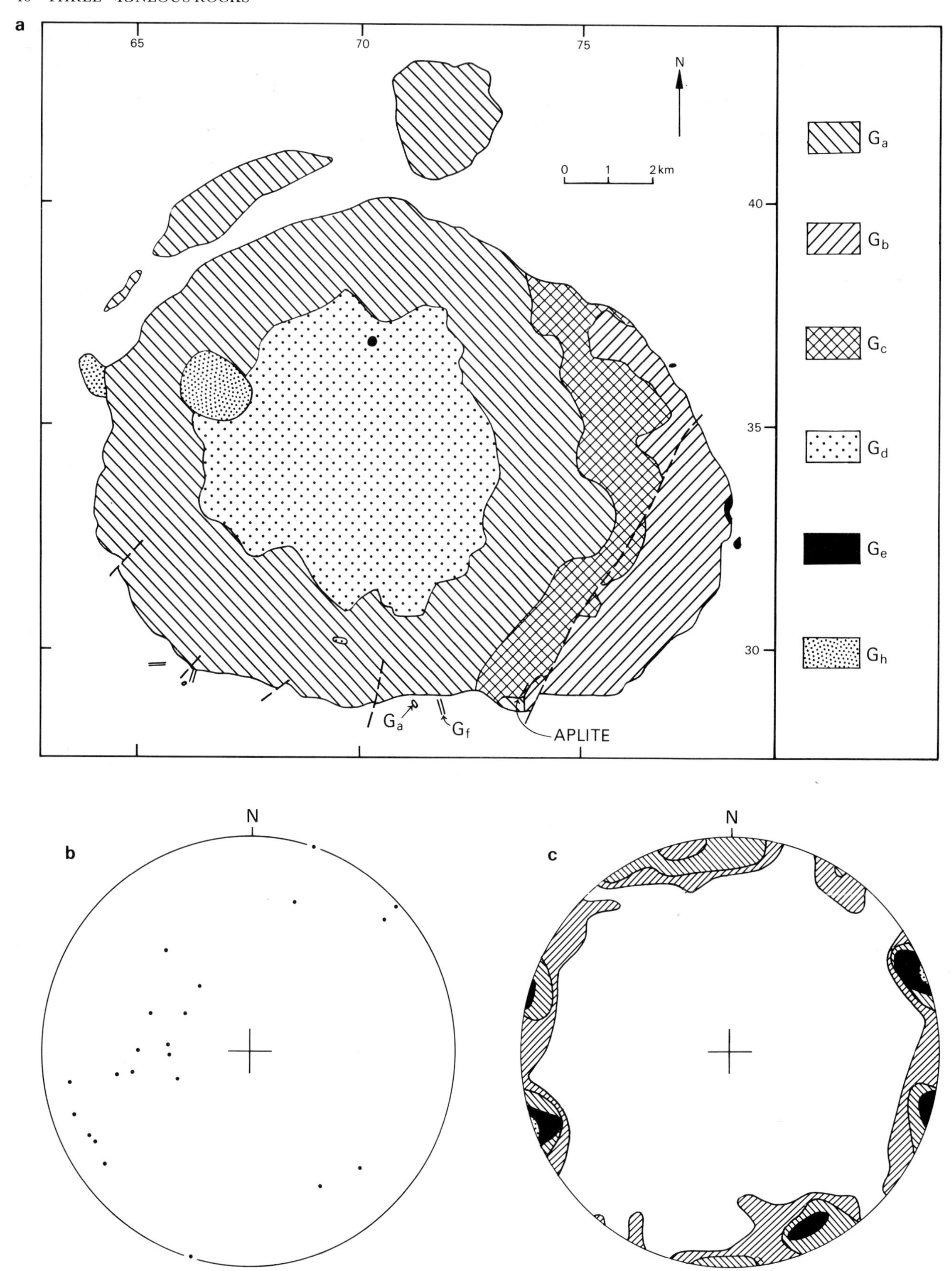

Figure 7 Granite structure: (a) sketch-map showing the textural subdivisions of the Carnmenellis Granite (b) 22 poles of feldspar megacryst alignments (c) 219 poles to joints; contours at 2%, 3%, 4%, 5%

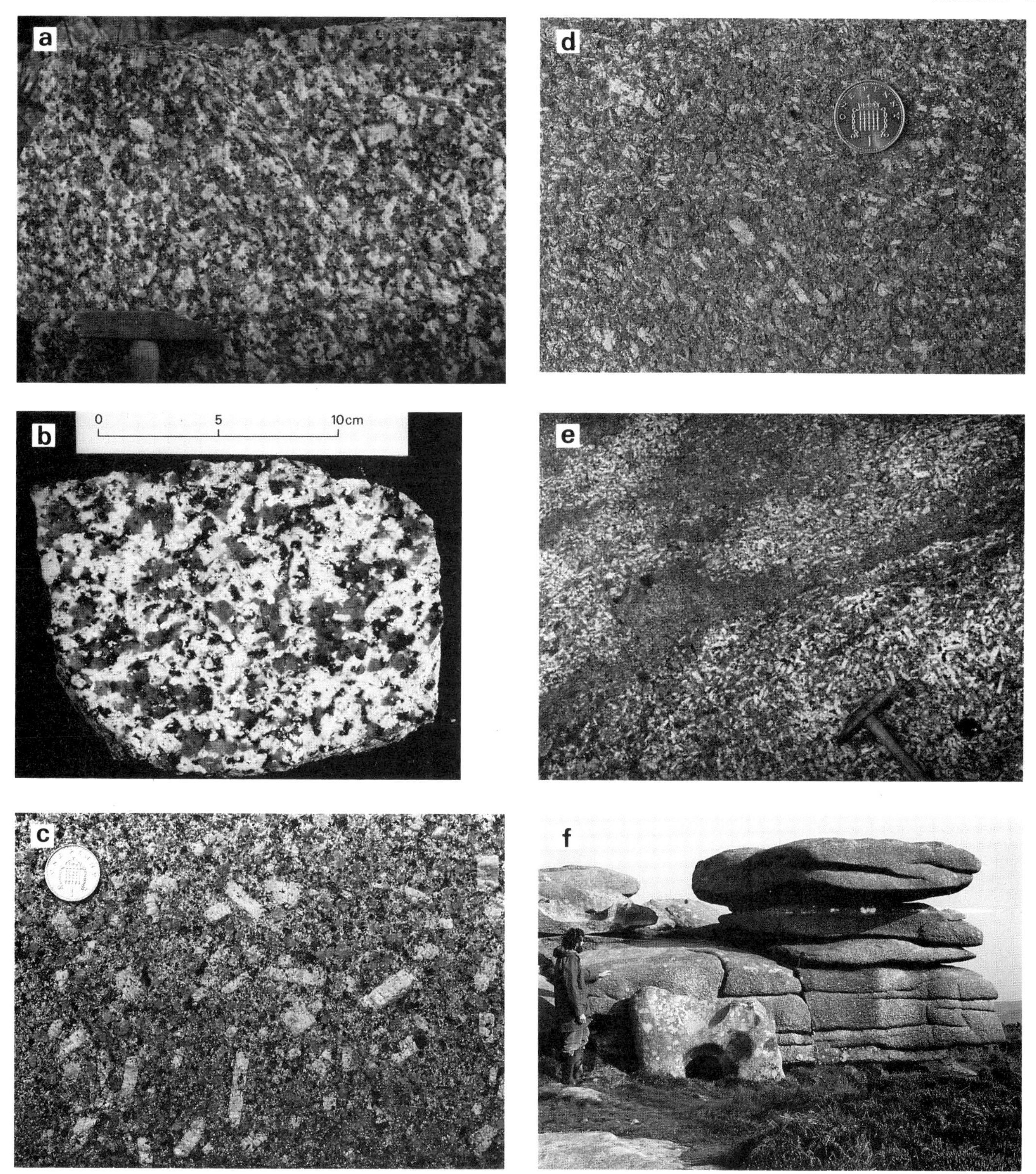

Plate 6 Granite textures, fabric and exposure

a) G_a texture. b) G_b texture. c) G_d texture. d) G_h texture. e) Mafic schlieren, Polkanuggo Quarry, Longdowns [7402 3467]. f) Tor on Carn Brea [683 407] (A J J Goode).

Trevone quarry [747 322]. They display sharp margins with the granite and are composed of an aggregate of biotite (80 per cent) with andalusite, quartz, tourmaline and accessory corundum. The rock exhibits a crude mineralogical banding defined by variations in the relative abundance of biotite, andalusite and quartz. Biotite forms 0.2–0.3 mm brown, pleochroic, subequant flakes with few pleochroic haloes and replacive microcrystalline white mica intergrowths around their edges. Small opaque grains, euhedral andalusite crystals, and corundum occur as inclusions. Some andalusite laths up to 1 mm long appear to overprint biotite. Less commonly, the mineral forms elongate aggregates of ragged grains up to 4 mm in length. They exhibit a faint pink to colourless pleochroism and contain abundant small opaque inclusions. Clear grains of quartz 0.2–0.4 mm in size, with a mosaic texture make up 15 per cent of the rock in bands with a low andalusite content. Yellow to blue-green, pleochroic tourmaline is also present forming anhedral crystals 0.2 mm in size in biotite-rich bands.

G_b This is a coarse-grained, equigranular to sub-megacrystic granite similar to G_a but contains only sparse, subhedral, white alkali feldspar megacrysts with a mean length of 19 mm (Plate 6b). It is a subdivision of Ghosh's (1934) Type 2, cropping out along the eastern margin of the pluton, and is partly in faulted contact with the G_c variety exposed inside the outcrop of G_b (Figure 7a). The G_b granite consists of alkali feldspar, plagioclase, quartz, biotite, white mica, chlorite and tourmaline (e.g [E58077, E58078]). Alkali feldspar textures are essentially similar to those seen in G_a, but lack the preferred alignments, plagioclase replacement and the distinctive quartz selvedges. In addition, in G_b some megacrysts display multiple zoning. Rounded aggregates of quartz, up to 10 mm in diameter, form interstitial growths between feldspars and partly replace alkali feldspar. Plagioclase forms sericitised 2 mm euhedral to subhedral crystals with multiple twinning. As in G_a it forms aggregates associated with biotite, which it partially encloses. Biotite shows the same textural features in G_a and G_b. However, apatite inclusions are less common and white mica and rare yellow pleochroic tourmaline are present as alteration products in addition to chlorite and epidote.

G_c This coarse-grained granite contains abundant white alkali feldspar megacrysts, generally less than 15 mm, displaying a preferred alignment in some places. It forms a subdivision of Ghosh's (1934) Type 2 granite and crops out in a 1–2 km wide arcuate strip between the G_a and G_b varieties in the east of the pluton, generally to the west of a north-north-east–south-south-west-trending fault (Figure 7). The G_c variety consists of alkali feldspar, plagioclase, quartz, biotite, white mice ± tourmaline (e.g. [E56706, E59882]). Alkali feldspar in G_c lacks a quartz selvedge to the megacrysts, but otherwise displays the same textural features as G_a. Quartz, interstitial with respect to biotite and feldspars, occurs as 3–4 mm pools between the feldspar/biotite aggregates. Plagioclase in G_c is similar to that in G_a but the 2–4 mm laths are only subhedral in the former and biotite may partially enclose plagioclase as well as vice versa. Biotite is also texturally similar in G_a and G_c. However, in the latter, single 4 mm flakes are present as well as aggregates, and white mica appears to be the only replacement mineral.

Dark xenoliths occur within the G_c granite and are composed of approximately 70 per cent biotite and 30 per cent plagioclase [E56701, E56703]. Aggregates of brown pleochroic biotite flakes up to 4 mm in length with abundant pleochroic haloes occur close to the xenolith contact, intergrown with largely untwinned 0.2–0.4 mm anhedral plagioclase grains. Sporadic, pale brown pleochroic tourmaline in euhedral prisms up to 3 mm long is present near the contact of some xenoliths.

G_d This variety is entirely equivalent to Ghosh's (1934) Type 3 granite and crops out as a broadly circular mass in the centre of the pluton (Figure 7a). It is easily distinguished from the other granite phases by its generally medium-grained matrix (about 1 mm) and abundant alkali feldspar megacrysts, with a mean length of 21 mm showing no preferred alignments (Plate 6c). The matrix grain-size is not constant, however, and irregular patches of coarser- and finer-grained granite occur within the outcrop. Texturally [E59884, E59911], alkali feldspar in G_d shows the same features as in G_a. Quartz forms large pools 4 mm in diameter interstitial with respect to the feldspars and biotite. In most sections it replaces biotite and feldspars both as small, round, seived inclusions and forming myrmekite. Larger quartz grains enclose corroded remnants of plagioclase, alkali feldspar and biotite. The 3 mm stubby laths of plagioclase, commonly in 10 mm aggregates, are texturally similar to the euhedral laths in G_a. However, they also show partial replacement by white mica and some crystals are strongly hematised. Biotite in G_d is similar to that in G_a, but also occurs as 3 mm single flakes and is commonly partially or totally replaced by white mica and less commonly by chlorite and epidote.

G_e and G_f The G_e and G_f varieties are both medium-grained biotite-muscovite-granite but they are differentiated by the presence of stubby, subhedral feldspar megacrysts up to 10 mm in length in G_f. They occur in widely distributed, small satellite stocks and dykes of variable composition around the margins of the Carnmenellis pluton, and as rare stock-like bodies within it (Figure 7a). In some intrusions, there are irregular patches and streaks of biotite schlieren. In places, notably at Kergilliack Farm [783 333], these rocks have been strongly kaolinised and are composed of kaolin, quartz and white mica. Both varieties consist of subequigranular aggregates of plagioclase, alkali feldspar, quartz, white mica, biotite, chlorite and tourmaline strongly altered by extensive quartz replacement.

Anhedra of perthitic alkali feldspar (0.2–0.4 mm) are present in the groundmass but, unlike other granite varieties, the feldspars are not clustered, but are evenly distributed throughout the groundmass as single grains [E58071]. Quartz is ubiquitous as inclusion-free, 0.2 mm anhedral grains intergrown with, and corroding, the feldspars and micas, and as very small round inclusions of replacement origin. Plagioclase is present both as 2 mm-long multiply twinned and sericitised subhedral laths with round quartz inclusions, and as 0.2 mm anhedral grains. Brown pleochroic biotite, with few pleochroic haloes, occurs as sporadic ragged flakes corroded and sieved by small quartz grains and strongly altered to chlorite and white mica. Zoned yellow pleochroic tourmaline in anhedral grains, and ragged 2 mm white mica flakes replacing alkali feldspar and plagioclase are widespread. In the G_f granite the 6–10 mm perthitic alkali feldspar megacrysts are subhedral with 0.6 mm zoned euhedral plagioclase inclusions and a selvedge of small round quartz grains.

G_h This granite corresponds to the fine-grained granite of Ghosh (1934). It occurs in two masses with subcircular outcrop patterns, the larger approximately 2 km in diameter straddling the G_a/G_d contact at Boswyn [665 366] and the smaller, less than 1 km in diameter, on the pluton contact at Praze-an-Beeble [640 360]. The rock types in both these masses have a mean matrix grain-size of approximately 0.5 mm, but show a considerable variety of composition and texture. Commonly, the rock is leucocratic and equigranular to submegacrystic with scattered white to buff alkali feldspar laths up to 3 mm long in a fine-grained matrix of anhedral alkali feldspar, plagioclase and quartz with variable amounts of biotite and white mica (Plate 6d). Rare 10 mm alkali feldspar megacrysts may also be present. In some localities there are dark grey rounded tourmaline nodules 20–30 mm in diameter. This tourmaline, in contrast to that found elsewhere in the granite, is bluish green and zoned (Ghosh, 1934).

A distinctive variety of the G_h granite is represented around Crowan Beacon [664 352]. This variety is characterised by abundant slender euhedral white feldspar megacrysts up to 10 mm in length aligned within a planar flow foliation. The megacrysts are enclosed within a fine- to medium-grained matrix of alkali feldspar, plagioclase, white mica, biotite and quartz with abundant disseminated tourmaline. Granite similar to G_h has also been encountered at less than 2 km depth below the G_a granite in the geothermal borehole in Rosemanowas quarry [735 346].

Aplite A small number of aplite dykes, varying in thickness from 0.2 m to 10 m cross-cut the G_a and G_c granites. They are generally grey to buff, mesocratic, fine grained with 10–12 mm stubby euhedral alkali feldspar and 5 mm quartz megacrysts [E58080]. The groundmass is subequigranular with euhedral to subhedral alkali feldspar megacrysts and 0.3 mm euhedra in the groundmass. Alkali feldspar is perthitic, twinned, includes small quartz grains and is partially replaced by white mica. Quartz is present in aggregates up to 3 mm in diameter, interstitial to alkali feldspar. White mica is abundant as 0.4 mm ragged flakes and 1 mm aggregates partially enclosed by quartz, but interstitial to alkali feldspar. Tourmaline, as small yellow pleochroic subhedral prismatic crystals, is present replacing alkali feldspar and white mica.

Schlieren Mafic schlieren are not common in the Carnmenellis Granite but do occur at several localities associated with the G_a and G_c granites.

At Polkanuggo Quarry [743 345] a mesocratic, medium-grained, equigranular granitoid, with a relatively high proportion of biotite occurs as irregular wispy schlieren less than 0.15 m thick (Plate 6e). These schlieren trend about 340°N and dip eastwards between 30° and 80°. In general their upper contacts with the G_a granite are sharp, but their lower contacts are always gradational. Loose blocks at Carnsew Quarry [762 346] display 0.05 m thick feldspar-rich pegmatite veins along the G_a/G_c contact. Laminar biotite schlieren lie parallel to this boundary and are interlaminated with G_c granite that locally shows strong feldspar alignments parallel to the schlieren. Similar feldspar rich pegmatite veins in the G_c granite south-west of Carvedras [7367 3066] are associated on one margin only with an irregular zone of nonmegacrystic biotite-rich granite up to 0.3 m in thickness.

Petrogenesis The origin of megacrysts of alkali feldspar within the south-west England granites has been the subject for some controversy. Ghosh (1934) described them as magmatic phenocrysts. Exley and Stone (1964) suggested they were late subsolidus features produced by potassium metasomatism of an originally aplitic textured granite. Hawkes (1967, 1968) proposed that they were replacements of earlier plagioclase phenocrysts and Stone (1979) suggested a subsolidus metasomatic origin.

Textural evidence obtained during the present study suggests a magmatic origin for megacrysts in the Carnmenellis Granite. Megacrysts of alkali feldspar, apparently growing across the contact of an aplite vein cutting an elvan, have been noted in a borehole at Carwynnen. In detail these feldspars are not single crystals but are divided into two parts by a line of inclusions defining the contact between the vein and host rock. Both quartz and feldspar show similar features suggesting that the portion of the crystals within the aplite nucleated and grew from the magma in the vein in optical continuity with the wall-rock crystals as a syntaxial overgrowth (Ramsay, 1980). No examples of "rapakivi" texture (cf. Hawkes, 1967) have been noted in the Carnmenellis Granite, but some of the plagioclase included in alkali feldspar megacrysts is in optical continuity with perthitic laminae suggesting a common origin by exsolution. Other plagioclase inclusions are euhedral and unrelated to the perthitic lamellae. As there is no evidence of corrosion of these inclusions the proposals of Exley and Stone (1964) and Hawkes (1967, 1968) that they represent the remains of partially replaced plagioclase appears unsupported. Both the plagioclase and biotite inclusions are smaller within the megacrysts than they are in the groundmass implying that the megacrysts began to crystallise before crystallisation of these minerals had ceased. However, anhedral alkali feldspar within the groundmass has no inclusions implying commencement of crystallisation later than the megacrysts. Small, rounded quartz grains are present with larger, irregular-shaped quartz crystals in the megacrysts. In sections showing strong quartz replacement textures, similar quartz grains are also present in plagioclase and biotite crystals, implying that this quartz has a replacement origin. This is supported by the occurrence of quartz selvedges around megacrysts in the G_a granite. The small quartz grains are an order of magnitude smaller than quartz in the groundmass and are unlikely, therefore, to represent quartz entrapped by alkali feldspar overgrowing the groundmass, as suggested by Exley and Stone (1964). Furthermore no evidence has been found of alkali feldspar replacing

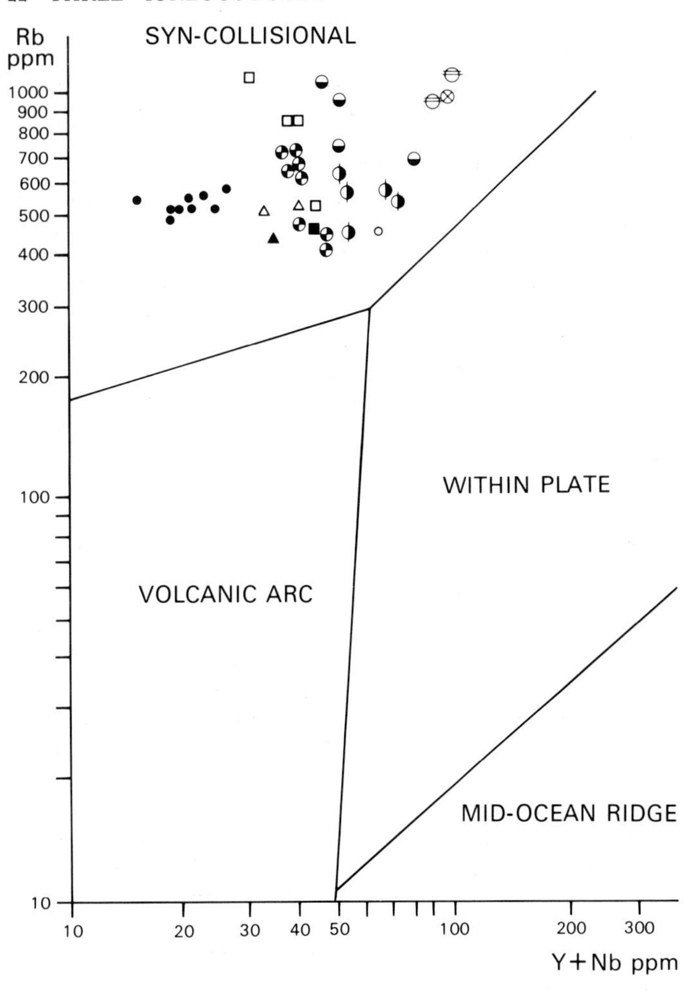

Figure 8 Rb vs (Nb + Y) discriminant plots for south-west England granitoids

plagioclase, altough the reverse is common. The alkali feldspar megacrysts therefore appear to represent true magmatic phenocrysts as described by Ghosh (1934).

The order of crystallisation of minerals in the granite varieties shows little variation. Biotite generally precedes plagioclase, although the reverse relationship is observed in the G_a granite. Alkali feldspar postdates plagioclase. Quartz, which in the groundmass partially encloses alkali feldspar, crystallised last.

The Carnmenellis Granite (Table 1) is peraluminous with 2–4 per cent normative corundum, a high oxygen value ($d^{18}O$ = 10.8–13.2 per cent; Sheppard, 1977) and a high initial $^{87}Sr/^{86}Sr$ ratio (0.7130 ± 0.0040; Darbyshire and Shepherd, 1985). These support the view that the magma possessed a large 'S-type' component (Brammall and Harwood, 1932; Exley and Stone, 1964; Exley et al., 1983; see also Chappel and White, 1974) and originated by lower crustal melting or was affected to a high degree by crustal contamination. Charoy (1986) demonstrated extensive hydrothermal reworking and proposed that the granite formed by crustal melting at about 800°C and 5 kb to form magma with 4 per cent water. The proposal of Exley and Stone (1964) that biotite is xenolithic, and Lister's (1984) suggestion that the rarity of xenoliths in the granite reflects their advanced assimilation are refuted by Jefferies (1985) who pointed out that contamination of the granite by pelitic xenoliths should cause elevated levels of Zr, Th and REE in the granite and that these had not been observed.

Xenolith assimilation therefore appears only to have had a minor influence on the granite chemistry and the sporadic occurrence of biotite with andalusite may be due only to a localised oversaturation of Al and Fe/Mg resulting from chance contamination.

Statistical analysis of the available data by Stone and Exley (1978) shows that Carnmenellis appears to contain less K and total alkalis, but more Na than other plutons in south-west England. Geochemical studies by Exley et al. (1983) and Alderton et al. (1980) indicate a high degree of magma fractionation, in the granite. The Rb-(Y + Nb) discriminant diagram of Pearce et al. (1984) was in part based on the chemistry of the south-west England granites but Figure 8 serves to illustrate the position of the Carnmenellis pluton in relation to other south-west granitoids in the syncollisional field.

Geochronology Miller and Mohr (1964) measured K/Ar mica ages for the Bodmin, Carnmenellis and St Austell granites that all clustered around 286Ma. Rb/Sr whole rock

Table 1 Chemical analyses of granitic rocks in the Falmouth district

	G_a per cent	G_c	G_d	G_h	Elvan
SiO_2	72.27	72.24	73.06	73.89	72.90
Al_2O_3	15.51	15.59	14.46	14.62	13.83
Fe_2O_3	2.10	2.16	0.43	0.32	0.47
FeO	—	—	1.36	0.91	1.18
MgO	0.58	0.58	0.03	0.32	0.31
CaO	0.99	0.96	1.06	0.33	0.72
TiO_2	0.20	0.22	0.16	0.15	0.17
Na_2O	3.07	3.10	3.72	0.32	0.69
K_2O	5.28	5.14	4.70	8.11	7.22
H_2O	—	—	1.00	1.01	1.83
TOTAL	100.00	99.99	99.98	99.98	99.61
Trace elements are given in units of µg/g					
Ba	174	165	—	—	130
Cu	—	—	—	—	10
Ni	3	4	—	—	10
Sr	97	97	—	—	56
Zr	102	113	—	—	20
As	15	37	—	—	—
Rb	521	512	—	—	—
Y	9	8	—	—	—
Sn	12	17	—	—	—
Ce	60	39	—	—	—
Pb	38	39	—	—	—
U	14	14	—	—	—
Zn	34	48	—	—	—
Nb	10	12	—	—	—
Co	—	—	—	—	10
Cr	—	—	—	—	10
Ga	—	—	—	—	15
Li	—	—	—	—	82
Sr	—	—	—	—	56
V	—	—	—	—	10
B	—	—	—	—	200
F	—	—	—	—	3300
S	—	—	—	—	30

G_a granite, Trevone Quarry [747 322]
G_c granite, Carnsew Quarry [758 344]
G_d granite, Carnmenellis Hill [695 364] from Ghosh (1934)
G_h granite, East part of Bolitho–Boswyn mass from Ghosh (1934)
Elvan, South Crofty (Hawkes et al., 1975)

data determined by Darbyshire and Shepherd (1985), indicate a consolidation age of 285 ± 9 Ma. Their mineral data from alkali feldspar, plagioclase, biotite and muscovite showed a range of Rb/Sr ages from 281 ± 3 Ma to 290 ± 2 Ma. They reject the youngest age and suggest that 290 ± 2 Ma best reflects the age of consolidation of the Carnmenellis Granite. Charoy (1986) also noted the difference between Rb/Sr whole rock and mineral ages, quoting a whole rock age of 259 ± 5 Ma and an alkali feldspar age of 295 ± 15 Ma. Rb/Sr age data for postintrusion mineralisation around the granite give 269 ± 4 Ma for main stage mineralisation in South Crofty.

Structure Using the regional Bouguer Gravity Anomaly map (IGS, 1975) the dip of the granite/country rock contact to the south and east of the granite appears steep. In the south-west and north, however, the contact is shallowly dipping and, there is sub-surface connection between the main granite, the Tregonning Granite off the sheet to the south-west, and the satellite masses of Carn Marth and Carn Brea to the north. Gravity modelling (Al-Rawi, 1980) and seismic refraction studies (Brooks et al., 1984) of a north–south section through the granite suggest that it has a tabular form with a flat base at 10 km depth.

The evidence of the form of the intrusion from internal contacts in the granite is ambiguous. The G_b/G_c contact is nowhere exposed although for much of its length it is marked by a steep fault. Because this fault has only a small displacement of the pluton contact its control of the G_b/G_c contact implies that this is shallowly inclined. The contact between the G_c and G_a granites is transitional where it is exposed in Carnsew and Trelubbas quarries, but because it is poorly defined elsewhere its outcrop pattern provides no clue to its overall geometry. In contrast the contact between G_a and G_d is well defined and its strongly reticulate outcrop, independant of topography, suggests control by steep joints. The G_h granite generally occupies high ground and, from its outcrop pattern, may represent a sheet-like body intruded into the G_a granite.

Mineral alignments Both the G_a and the G_c varieties show a preferred orientation of alkali feldspar megacrysts defining linear and/or planar fabrics (Figure 7b) although the groundmass minerals show no alignments or signs of deformation. In orientation the mineral alignments define a loose group with a trend between 330° and 350°N, dipping between 40° and 80° to the north-east. Although this trend is broadly coincident with that of the major north-north-west joints the lack of any clear association between the alignments and the proximity or frequency of jointing, and the apparent magmatic origin of the megacrysts precludes a causative relationship.

The megacryst alignments are sporadically developed and extend laterally for a few tens of metres only, terminating in a swirling pattern in which the preferred alignment becomes progressively weaker and diffuse. Across the fabric they show a rapid transition to unaligned megacrysts. Typically, even where megacrysts are strongly aligned, some deviate from the fabric, and in places lie at right angles to it. This relationship, together with the lack of deformation, suggests that the alignments do not have a tectonic origin but are the products of convection current activity within the Carnmenellis Granite. Calculations of the viscosity and the estimated convection current velocities of the Ga and Gc granites, based on the work of Shaw (1972), are presented in Appendix 3.

Jointing Four main joint trends are present in the granite (Figure 7c):

1 Vertical joints trending 340°N are laterally extensive and, in places, carry purple flourspar and galena. They are parallel to the crosscourses present in the country rocks.

2 Inclined joints trending 242°N and dipping 70° to 80°

are also laterally extensive and commonly carry tourmaline and quartz coatings. In places, they are striated, indicating oblique-slip movement on the joints. These joints are parallel to the mineral lodes and elvan dykes in the country rocks and in the granite.

3 Vertical joints trending 012°N are sporadically developed, unmineralised joints with no counterparts in the country rocks.

4 Vertical east–west-trending joints are also sporadically developed and without counterparts in the country rocks.

The parallelism of the 340°N and 242°N joints with those in the country rocks, the pre-elvan and premineralisation age of these joints, and the lack of any radial or concentric cooling joints implies that the granite consolidated and was jointed during the imposition of a regional stress field, with the principal extension in the south-east/north-west quadrants.

Metamorphic aureole The subsurface shape of the Carnmenellis Granite is reflected in the shape of the metamorphic aureole. To the north of the granite the aureole is wide and extends as far as the coast north of Camborne and Redruth. On the south and east, where the granite contact is steep, the aureole rarely exceeds 1 km in width. The aureole is characterised by spotting of some of the slates. On the basis of the typical mineral assemblages it can be divided into outer and inner parts.

In the outer aureole, spots are composed of chlorite and/or white mica recrystallised from the groundmass, and no new mineral phases are present. In places the spots are deformed by the S_5 crenulation cleavage (see section four) preserving an earlier cleavage (?S_1) at a high angle to the crenulation cleavage.

The inner aureole is characterised by an assemblage of andalusite, biotite and white mica in pelitic rocks. The chlorite and white mica spots of the outer aureole are replaced in this zone by brown pleochroic biotite and white mica in a hornfels texture. In places, these have been deformed in the S_5 crenulation cleavage, but in the deformed spots opaque inclusions within the minerals preserve the trace of an earlier (?S_1) penetrative cleavage (Plate 7a, b). Where the S_5 crenulation cleavage is present in pelitic rocks it is defined by the alignment of biotite and white mica flakes. Where it is absent these minerals form reticulate or 'criss-cross' static overgrowths. Andalusite displays a hornfels texture and is present in three forms: ragged inclusion-filled porphyroblasts pseudomorphing the earlier mica spots, very large ragged aggregates overgrowing the penetrative cleavage, or euhedral prismatic crystals of chiastolitic andalusite. The last crystals are commonly zoned either with abundant opaque inclusions in the core and an inclusion-free mantle, or vice versa. Rarely, andalusite shows signs of minor deformation in the form of internal twinning or fracturing.

Andalusite is absent from silty rocks in the inner aureole and the mineral assemblage is limited to quartz, white mica ± biotite, with a strong alignment of micas defining the penetrative cleavage. Reticulate white mica and biotite static overgrowths are commonly superimposed on this fabric.-Sandstone in this zone shows little modification.

The assemblage of andalusite, biotite and white mica represents the highest grade of metamorphism in the Carnmenellis aureole (Hill and MacAlister, 1906). However, xenoliths within the Carnmenellis Granite from depths below the present exposure level record higher-grade assemblages that, according to Jefferies (1985), exhibit the following parageneses:

i A biotite-muscovite-sillimanite assemblage; the micas define a schistosity and sillimanite defines microfolds, possibly mimetically.

ii A spinel-corundum-ilmenite assemblage superimposed on the earlier fabric with a hornfels texture. The spinel is deep green with abundant corundum inclusions. Corundum encloses sillimanite needles and overgrows biotite. Monazite porphyroblasts overgrow sillimanite in the outer zones of the xenoliths.

iii An andalusite-biotite-ilmenite +/- cordierite assemblage. This frequently completely replaces the earlier assemblages but in some xenoliths it is limited to the outer zone. Andalusite is pink and pleochroic and mantles spinel and corundum. Biotite forms a hornfels texture. Cordierite, extensively pinitised, mantles the xenolith.

According to Jefferies (1985) assemblages i) and ii) represent prograde metamorphism of the granite wall rocks, the second being associated with desilication and alkali loss. Assemblage iii) reflects gains in silica and alkalis during reaction between the granite and xenoliths after their inclusions.

In summary, the metamorphic textures indicate a four-stage sequence:

i Formation of chlorite and white mica spotting throughout the aureole.

ii Sporadic deformation of thermal spots by the S_5 crenulation cleavage.

iii Formation of an andalusite-biotite-white mica assemblage in the country rocks and a sillimanite-biotite-muscovite assemblage in some xenoliths, both superimposed on the thermal spotting.

iv In xenoliths only the formation of spinel-corundum-ilmenite assemblages with hornfels texture superimposed on iii).

v Formation of an andalusite-biotite-ilmenite-cordierite assemblage overgrowing iv).

PORPHYRY DYKES (ELVANS)

An extensive suite of quartz-feldspar porphyry dykes traverses both the granite and country rocks in south Cornwall. The dykes commonly occupy steep fractures and vary in width between a few tens of centimetres to more than 40 m. In general the dykes trend east-north-east–west-south-west. Some east–west-trending dykes are present locally, and there is a zone of north–south-trending dykes between Carnon Downs and Mylor Bridge. Despite variations in the trends of elvan dykes they are locally always parallel to the hydrothermal mineral veins and are, in places, associated with breccia dykes (Goode, 1973; Goode and Taylor, 1980). In the Falmouth district elvan dykes are common in the Mylor Slate Formation and the granite, especially in the mineralised zone around Camborne and Redruth,

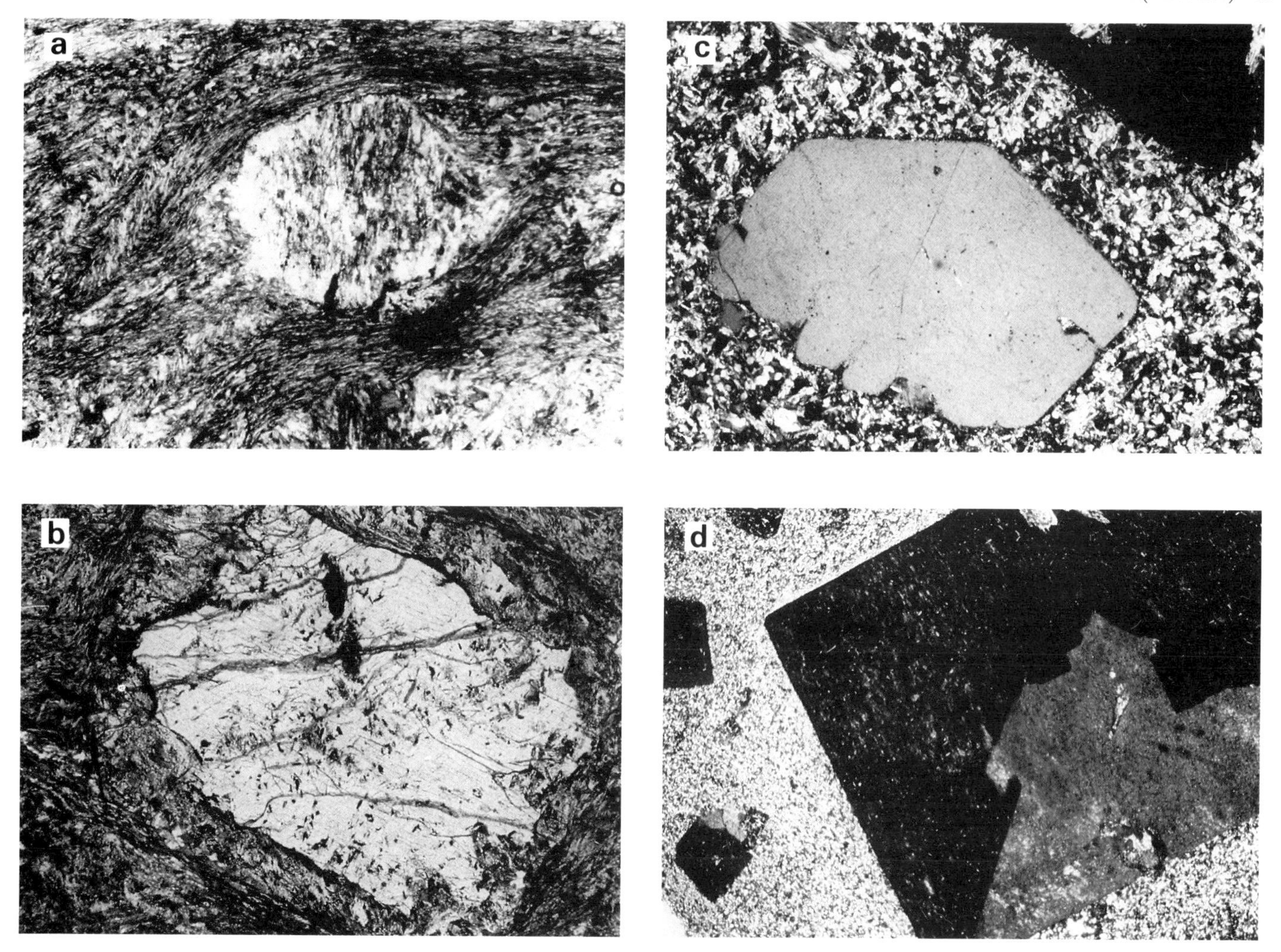

Plate 7 Photomicrographs illustrating the fabrics and mineralogy of the metamorphic aureole of the Carnmenellis Granite (a,b,) and phenocryst and matrix textures of the elvans (c,d).

All ×50. a) Strongly deformed andalusite porphyroblast showing rotation of S_1, recorded by inclusion trails, by the ?S_5 crenulation cleavage [E56666]. b) Zoned euhedral andalusite porphyroblast with a chiastolitic inclusion cross in the core and static overgrowth of the ?S_5 crenulation in the surrounding mantle [E56667]. c) Corroded quartz phenocryst (centre) with an alkali feldspar phenocryst (top right) and chlorite pseudomorphing a biotite phenocryst (top left) in a fine-grained matrix [E59899]. d) Euhedral perthitic feldspar phenocryst in a microcrystalline quartz-feldspar-white mica matrix [E59960].

and less common in the Porthtowan and Portscatho formations.

Petrography Typically the elvans contain 5–30 per cent phenocrysts (Henley, 1974), which, in order of abundance, are alkali feldspar, quartz and subordinate plagioclase and chlorite. Quartz and feldspar phenocrysts are commonly fringed by myrmekite.

Alkali feldspar forms euhedral perthitic phenocrysts up to 8mm in length, occurring as single crystals and monomineralic aggregates, or polymineralic aggregates with quartz and plagioclase (e.g. [E59894, E59903]). Both chlorite flakes and plagioclase are wholly or partially enclosed by alkali feldspar. In an elvan at Bolenowe Carn Moor [6705 3738] [E59900] alkali feldspars have been preferentially replaced by blue pleochroic tourmaline.

Quartz is present as corroded euhedral to subhedral, rounded 2–5 mm phenocrysts forming single crystals (Plate 7c), monomineralic aggregates and, less comonly, aggregates with feldspars. At Nine Maidens Downs [E59902] smaller quartz phenocrysts (0.2–0.3 mm) are also present. In the polymineralic aggregates quartz is commonly graphically intergrown with the feldspar. Inclusions of small plagioclase crystals and brown pleochroic biotite, or its pseudomorphs occur sporadically. Plagioclase phenocrysts form clusters of euhedral, sericitised, or partly chloritised, multiply twinned crystals, 0.5 to 5 mm long, or polymineralic aggregates with alkali feldspar ± quartz. The phenocrysts enclose rare aggregates of chlorite and opaque minerals pseudomorphing biotite. The groundmass is composed of quartz and feldspars in a cryptocrystalline to microgranitic intergrowth with minor chlorite (Plate 7d).

Alteration is common and variable in intensity. Secondary minerals include white mica, blue or yellow pleochroic tourmaline, and brown or green biotite. In two dykes at Carwynnen [E59894] [658 366] and Croft Michael [E59898] [663 369] small brown biotite flakes surround and are partially enclosed within euhedral and anhedral, pink, pleochroic andalusite porphyroblasts. At Gwarnick [6415 3635] all the original igneous minerals and textures have been replaced and the rock is now represented by finely intergrown ragged porphyroblasts of quartz, white mica, chlorite and iron oxides.

The sequence of crystallisation of the elvans is biotite, plagioclase, quartz and alkali feldspar. However, the corrosion of quartz and feldspar phenocrysts suggests that these minerals were, to some extent, xenocrystic.

Petrogenesis Geochemically, elvans show a granitic composition, but with a trend towards potassium enrichment and impoverishment of sodium (Stone, 1968). Hawkes and others, (1975) suggested that this was a secondary effect. All analyses show oversaturation in alumina, with up to 12 per cent corundum appearing in the CIPW norm (Henley, 1974). Despite these differences there is some overlap between the range of compositions shown by elvans and those of the main granite phases of the south-west England batholith. Stone and Exley (1978) indicated that some elvan compositions clustered within the biotite-granite compositional field and others showed similarities with the fine-grained granite, G_h, of the Carnmenellis pluton. Darbyshire and Shepherd (1985) suggested similarities in Europium anomalies and LREE/HREE ratios between the South Crofty elvan and the Carnmenellis Granite, implying either a genetic link between granite and elvan or extensive contamination of the elvan magma by local granitic material.

Stone (1968) and Henley (1974) suggested that the elvans formed by modification of the main granite types, Stone (1968) by potassium ion exchange from a granite melt and Henley (1974) by a potassium-rich aqueous fluid. He postulated that this fluid formed during the digestion of pelitic xenoliths at deep levels in the granites; then it accumulated at the top of the consolidated pluton and caused remelting of the granite and the production of a potassium-rich elvan melt. The variable petrography and strontium isotope ratios of the elvans reported by Hawkes et al. (1975) led them to suggest heterogeneous source rocks.

Age of intrusion The elvan dykes intrude all parts of the Carnmenellis Granite, except the fine-grained (G_h) granite at Boswyn, and are themselves cut by the hydrothermal mineral veins. Dykes in the Carnmenellis Granite at Carwynnen are cross-cut by pegmatitic granite veins suggesting that elvan intrusion overlapped with the later stages of consolidation of the granites. According to Goode and Taylor (1980) the elvans are generally synchronous with intrusive breccias with which they may share a genetic relationship.

Radiometric dating of two dykes of the elvan swarm has given ages of 282 ± 6 Ma and 270 ± 3 Ma (Darbyshire and Shepherd, 1985) indicating a long history of formation.

Structure The marginal zones of many dykes differ from the central zones in being more silica-rich (Stone, 1968). They are almost devoid of phenocrysts and display a variety of flow structures described by Goode (1973). These include flow laminations parallel to the dyke walls, or in swirling patterns around irregular walls or large xenoliths. The laminations may be isoclinally folded, refolded folds, boudinaged or brecciated. Strings of spherulites are also present marking flow lineations. The fine-grained marginal zones commonly exhibit sharp contacts against a megacrystic central zone in which the megacrysts and platey xenoliths may be orientated parallel to the dyke walls and display an increase in size towards the centre of the dyke. These planar laminations were thought by Goode (1973) to be the result of a turbulent fluidised intrusion regime related to the formation of intrusive breccias, but it is more likely that they reflect the laminar flow of a viscous magma.

In planar-walled fissures, shear against the walls caused by the injection of the elvan magma creates a velocity gradient perpendicular to the dyke trend, with flow planes parallel to the walls. Inhomogeneities in the magma would be drawn out by shear into laminations parallel to the walls. Steps in the wall profile would produce perturbations in the flow that, with subsequent shearing, would create assymmetric drag folds in the flow laminations. Initially these folds would have axes perpendicular to the flow direction, but continued shear would cause the formation of sheath folds (Cobbold and Quinquis, 1980) that, as they were drawn past successive steps in the dyke wall, would become refolded (the 'eloborate swirls and eddies' of Goode (1973)). The megacrystic elvan present in the centre of some of these dykes seldom shows flow-orientation (Goode, 1973) and must therefore represent a more fluid magma capable of turbulent motion. This contrast in viscosity may be related to a higher partial pressure of water in the centre of the zoned elvans. Flow-banded elvans may therefore represent magmas in which volatiles have been lost, possibly due to a pressure drop as the elvan fissure connected to the surface. The volatiles would be available possibly to form intrusive breccias which could subsequently be intruded by the flow-banded magma. If this activity blocked the fissure, or if the fissure failed to connect to the surface, there would be no volatile loss and a low viscosity magma, thoroughly mixed by turbulent flow, would be intruded.

FOUR

Structure

The presence of isoclinal folds with limbs and related cleavage dipping to the south-east, crossed by a later strain-slip cleavage, was recorded during the previous survey of the district (see Hill, 1913). Dearman (1969) and Sanderson (1972) referred to early folds in the area as examples of folding oblique to regional trend, but a detailed structural analysis was not forthcoming until Leveridge (1974) described and attributed folds and related cleavage fabrics to three main phases of deformation in Roseland. Polyphase deformation was recognised elsewhere in south Cornwall by Lambert (1959) in Meneage, Stone (1962) in Mounts Bay, and Smith (1965) on the north coast at Godrevy, but there had been no consensus on the correlation of the structures. Subsequent work (Turner, 1968; Dearman et al., 1980) demonstrated that in some instances separable major deformation episodes had been overlooked. More recently, Rattey and Sanderson (1982, 1984) reported a structural sequence to the north of the Lizard between Carrick Roads and Mounts Bay that was in broad agreement with Turner (1968). They recognised N-S zones of 'regional' east-north-east-west-south-west-trending, north-facing folds alternating with north-south belts of oblique folds facing west. Oblique folding was attributed to north-west-south-east belts of sinistral shear related to differential advance of an overriding Lizard thrust sheet.

Strata in the Falmouth district of Cornwall have a prevailing dip to the south-east. They are predominantly the right way up and the lithostratigraphical divisions young towards the south-east. The rocks, however, have been strongly deformed.

Five major episodes of folding and related cleavages, several later minor phases of folding or cleavage development, and faulting both related to, and subsequent to the main deformations have been recognised during the recent survey. The correlation of the main deformation phases with previously described structural sequences is shown in Table 2.

Structural measurements are largely restricted to the coastal sections where there is sufficient exposure for fabric relationships to be observed. It is evident that D_1, the major deformation phase, is ubiquitous, whereas structures of succeeding phases, despite being locally intense, are only sporadically developed. This being so the limited dissection and poor accessibility of the steep north-east-south-west-trending north coast does not necessarily provide a representative section of post-D_1 structures. In order to evaluate fully the structure of these rocks and avoid an overwhelming bias towards structures on the well-exposed south coast, data from outside the the sheet area on the north coast had to be taken into consideration. It was found that D_2 decreases in intensity north-westwards, D_3 weakens south-eastwards, D_4 persists through the district and D_5 is patchily developed largely within the Mylor Slate Formation in the vicinity of the granites.

Table 2 Chronology and correlation of the main deformation phases in south Cornwall

BGS	Rattey & Sanderson (1982)	Leveridge (1974)	Turner (1968)	Smith (1965)	Lambert (1959)
D_1	F_1	F_1	F_1	—	—
D_2	F_2	F_2	F_{2b}	—	F_1
D_3	—	—	—	F_1	—
D_4	F_2	F_3	F_{2a}	F_2	F_2
D_5	F_3	absent	F_3	F_3	—

FIRST PHASE DEFORMATION (D_1)

S_1 Cleavage The early cleavage is a penetrative structure, generally parallel to bedding, in all of the bedded rocks. In pelitic rocks it is a slaty cleavage. The associated cleavage in sandstones is contiguous with slaty cleavage in fold hinges where it may show a slight fanning. Microscopically it is a combination of a preferred dimensional orientation fabric in microcrystalline matrix and broadly parallel pressure solution planes (Plate 8a). These are marked by a dark, finely divided residuum and locally by a fresh mica overprint. The intensity of pressure solution cleavage varies from discrete anastomosing discontinuities adjacent to larger grains to penetrative intergranular and even intragranular planes [E 59873]. Where cleavage is well developed dissolution of large grains and regrowth in strain shadows is apparent and some grains show pull-apart fabrics.

Cleavage in cherts is represented by a preferred dimensional orientation fabric of quartz and subordinate mica. Recrystallised carbonate in limestones similarly defines S_1. Widely spaced pressure solution planes are also present.

A grain lineation, due to the preferred orientation of the long axes of minerals in the plane of cleavage, is often apparent on S_1 surfaces. Trending south-south-east (Figure 12a) it is better developed on the south coast than the north. Strain shadow development, grain trails in the sandstones, and the long axes of both deformed clasts in microbreccia and radiolarian tests in chert of the Pendower Formation are parallel to this mineral lineation. Deformed 'concretions and pebbles' in coarse sandstone at Pendower show considerable extension parallel to grain lineation with X/Y values between 2.5 and 3 (Sanderson, 1972).

This extension is accompanied by flattening perpendicular to S_1. Axial ratios of ellipsoidal radiolarian tests of 7.2: 3.6:1 have been recorded by Leveridge (1974) on chert of the Pendower Formation. This places the deformation within the flattening field of the Flinn plot (Flinn, 1965).

The regional variation in S_1 strike and dip is indicated in Figure 9. It is apparent that the granite has affected S_1 orien-

Plate 8 D_1 and later structures

a) S_1 cleavage fabric in greywacke sandstone ×25 [E59957]. b) Hinge zone of an isoclinal fold in a sandstone-dominant sequence of the Portscatho Formation, showing a penetrative axial plane cleavage and six-fold thickening of beds [8897 3768]. c) S_2 crenulation cleavage banding crossed by a weaker S_4 crenulation cleavage ×30 [E59875].

b

a

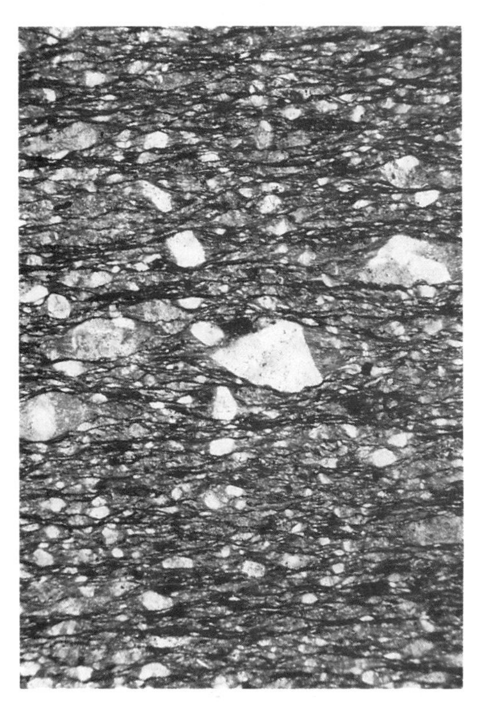

c

tation and attitude. The pattern is consistent with a north-easterly strike and gentle to moderate south-easterly dips prior to doming of the country rocks by the intruding granite, when dips on the north-west margin were decreased or locally reversed, and on the south-east side they were steepened. Elsewhere the strike was slightly modified to become subparallel to the granite margins except in opposing areas to north-east and south-west where accommodation structures with variable strike simulate macroscopic strain shadows.

Considering smaller subdomains, for example the 100 km^2 area of sheet SW 83, there are significant strike and dip variations up to 90° about the mode (Figure 10). This is accounted for by late open to gentle folds plunging gently to moderately between east and east-south-east, which produce a pole spread best illustrated in plot d in Figure 10. This is reflected on a larger scale by the strike change around the granite (plot a, 83NW and plot c, 83SW). Variation in dip is largely a function of D_4 structures and a further late phase of gentle folds trending north-east (see below).

F_1 folds S_1 is an axial plane cleavage and the bedding/cleavage intersection lineations are parallel to the

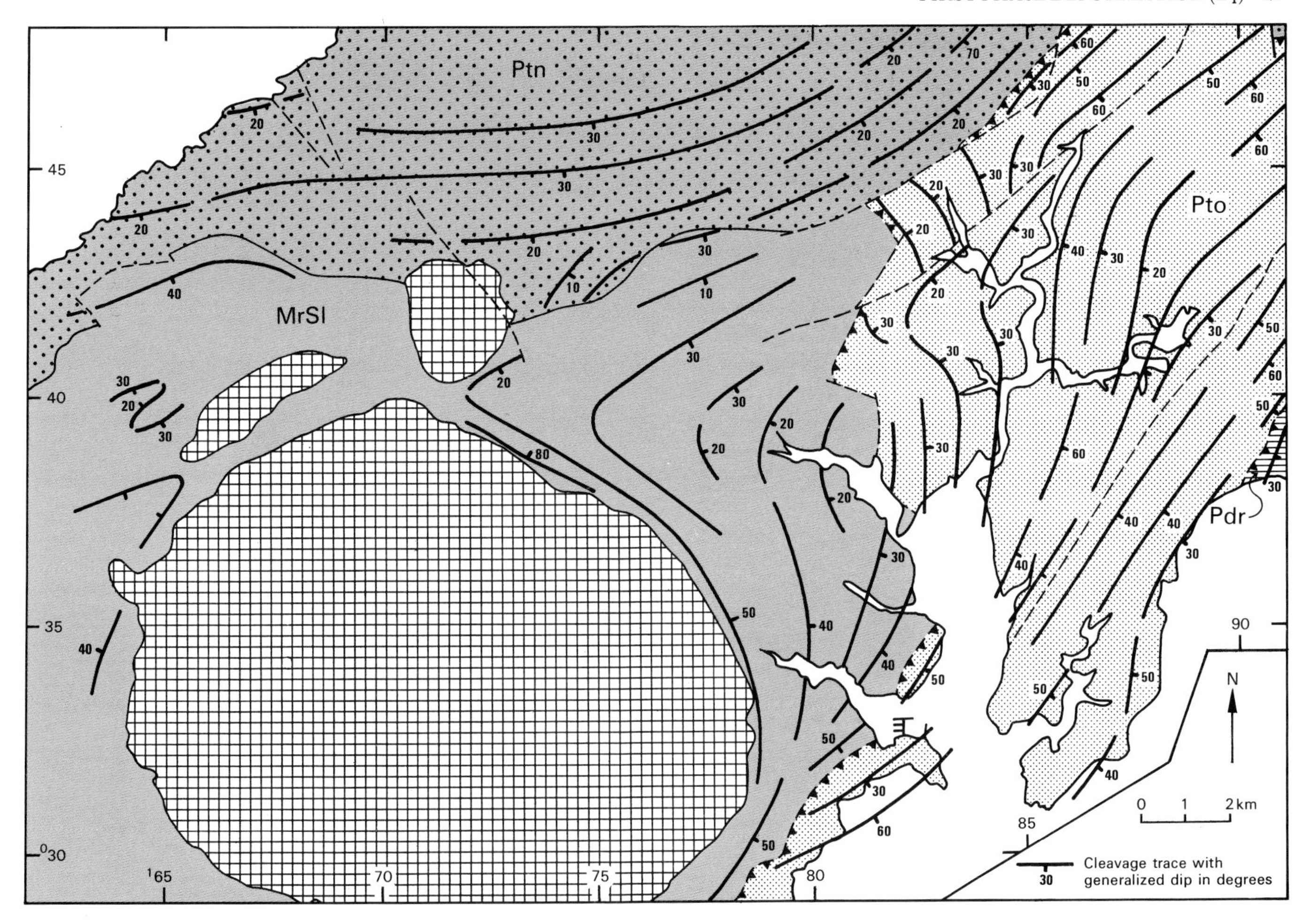

Figure 9 Sketch-map showing generalised strike and dip of S_1 in the Falmouth district

related fold axes. This, with the absence of inconsistent inversion and vergence or remanent earlier fabrics, indicates that F_1 and S_1 were coeval.

F_1 folds are asymmetrical structures with inverted shorter limbs. At most they affect only a few tens of metres of strata (e.g. [392 379]). They occur as isolated couplets or in families describing higher order folds.

Interlimb angles of the early folds in the sandstone facies rocks are generally in the close (e.g. [8865 3720, 8962 3805]) to tight (e.g. [8701 3218, 8899 3768]) range, although away from hinge zones limbs tend to be parallel to S_1. Exceptionally, folds in this facies are isoclinal near the hinge zone. In one such structure at Porthcurnick Beach [8795 3604] there is little hinge-zone thickening and delicate bedding structures are preserved. Interlimb angles of F_1 folds observed in the Mylor Slate Formation and in the chert and slate of the Pendower Formation tend to be very tight to isoclinal (Figure 11a). Fold hinge zones are commonly rounded, but some folds do have subangular or angular hinges. Angular hinges and constant interlimb angles produce a geometry akin to zig-zag folds except that cleavage is generally closer to the longer limb (e.g.[8881 3742]). F_1 fold profiles vary from parallel (type 1B of Ramsay, 1967) to those with geometry nearer similar folds (type 2 of Ramsay, 1967). Relative thickening of F_1 fold hinges by a factor of 6 is observed in folds between Pendower and Creek Stephen [8898 3769] (Plate 8b).

On the north coast F_1 folds are predominantly recumbent or gently inclined, plunging gently to the east or west (Figure 12c) and face northwards. In contrast, on the south coast, F_1 folds have variable attitude, pitch and trend (Figure 12b). On the scale of an outcrop individual folds may be inclined horizontal, inclined plunging or reclined, but throughout the area folds with the same trend can have any of these attitudes. Although there is a tendency for fold axes to be homoaxial within small subdomains, folds with different trends are juxtaposed at several localities, such as Pendower Hotel [8939 3797] (Figure 13) and near Creek Stephen [8885 3748]. Poles of F_1 fold axes and bedding/cleavage lineations, recorded largely along the coastal section, have a 360° spread, plotting mainly in the eastern hemisphere of the stereographic projection circle due to the prevailing southeasterly dip (Figure 12b).

Thrusts The major faults of the region are thrusts (Leveridge et al., 1984). Their presence is indicated by the bedded rocks consistently dipping and younging southeastwards, whereas determined palaeontological ages increase in that direction. The location of the faults at lithostratigraphical boundaries rather than within forma-

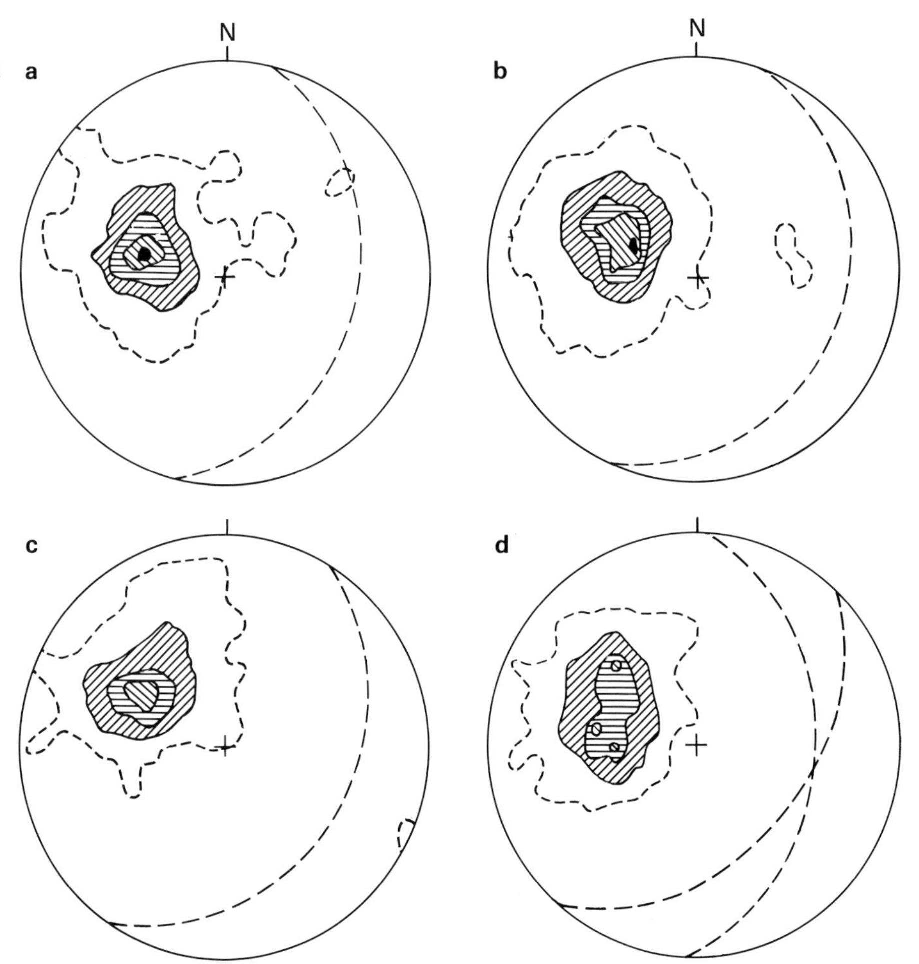

Figure 10 Contoured equal area projections of S_1 cleavage. Small dash lines represent limits of poles , larger dash lines are cyclographic traces of modal strikes and dips.

(a) SW 83 NW, 160 poles, intervals 5%, 10%, 15%, 20%
(b) SW 83 NE, 244 poles, intervals 5%, 10%, 12.5%, 15%
(c) SW 83 SW, 135 poles, intervals 5%, 10%, 20%
(d) SW 83 SE, 125 poles, intervals 5%, 10%, 12.5%

tions is consistent with the profound changes in sedimentary regime represented by those boundaries. In the sheet area there are two such faults, the Carrick Thrust (Leveridge, et al., 1984) and the Veryan Thrust. On land the Carrick Thrust is poorly exposed and commonly displaced by later faults. In the Falmouth area it follows regional strike and is parallel to bedding, but to the north of Restronguet the thrust passes, via a sidewall ramp, through the Mylor Slates and the underlying Porthtowan Formation to the top of the macroscopically folded Treworgans Sandstone Member. The thrust is exposed over a few metres of the Flushing foreshore [8165 3357] where folded and disrupted sandstone beds of the Portscatho Formation of the hanging wall are in sharp contact with the Porthleven Breccia Member. In the vicinity of the thrust the small siltstone clasts of the breccia member show extreme flattening but no appreciable directional elongation (Plate 5b). The extension of the Carrick Thrust offshore is well displayed in seismic sections and can be traced eastwards across Plymouth Bay and south-westwards across the South-Western Approaches (Day and Edwards, 1983).

A continuing supply of breccia clasts through the several hundred metres of the Porthleven Breccia Member required an active tectonic scarp as a source (cf. Cowan and Page, 1975). Leveridge and Holder (1985) have proposed that it was the emergent Carrick Thrust nappe, migrating over its own erosional debris. The presence of Portscatho clasts showing pre-incorporation tectonic deformation at Porthleven (Leveridge and Holder, 1985) and the similarity between D_1 fabrics in both the Mylor Slates and Portscatho Formation indicate that folding migrated north-westwards in association with thrust propagation. The presence of the Veryan Thrust is less certain than the Carrick Thrust. The upper parts of the Portscatho Formation are now established to be of Frasnian age. Although it is the right way up the overlying Pendower Formation contains Eifelian conodonts, which do not appear to be reworked assemblages, and which young upwards through the sequence (Sadler, 1973; Leveridge, 1974). The Veryan Thrust has therefore been proposed to account for the juxtaposition of the two formations. The junction is a moderately inclined fault where it is exposed above Pendower stream [9009 3859], but there is no obvious disruption at the boundary on the lower foreshore at Pendower [8970 3903]. The course of the fault suggests a contact conformable with dip of bedding and cleavage.

D_1 structure analysis

The distribution pattern of early folds recorded in the Roseland and Falmouth areas is largely a primary feature. Post-D_1 rotations can account for the variable facing sector of Figure 12b, but cannot alter F_1 trends and pitch angles by more than about 20° (Leveridge, 1974). The data do not support the concept that folds with a north–south trend facing

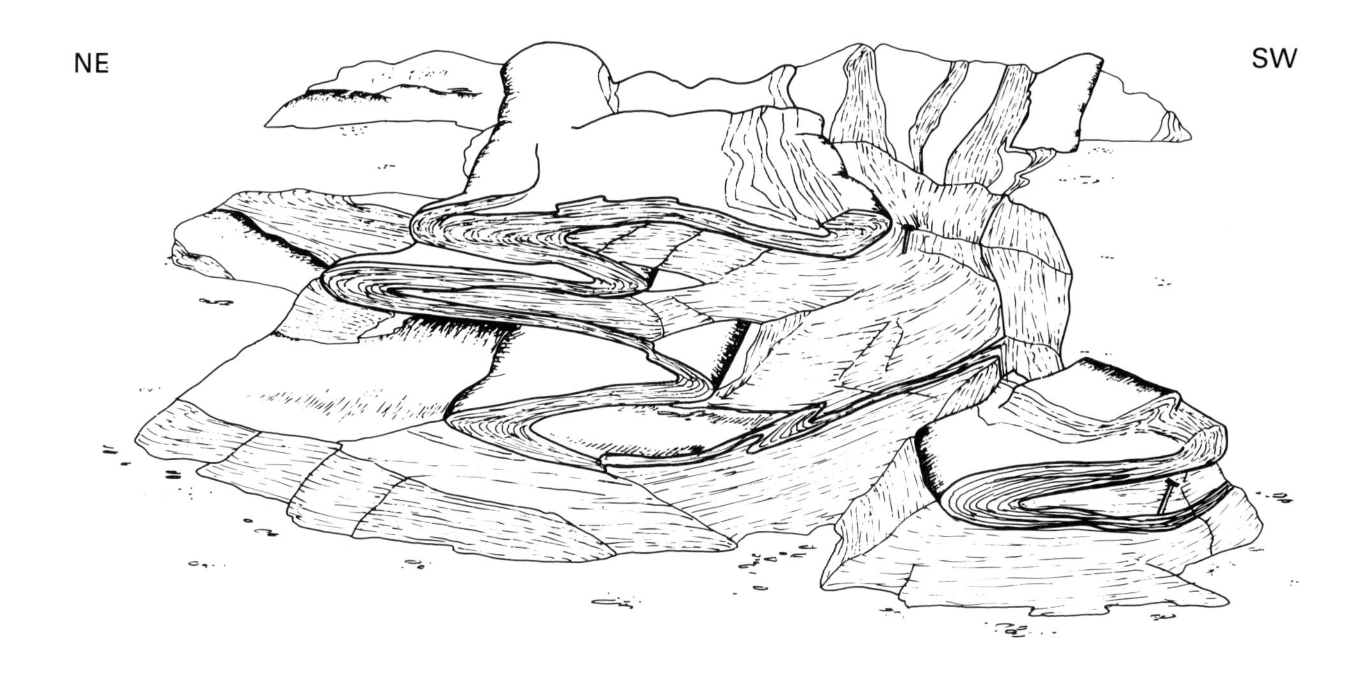

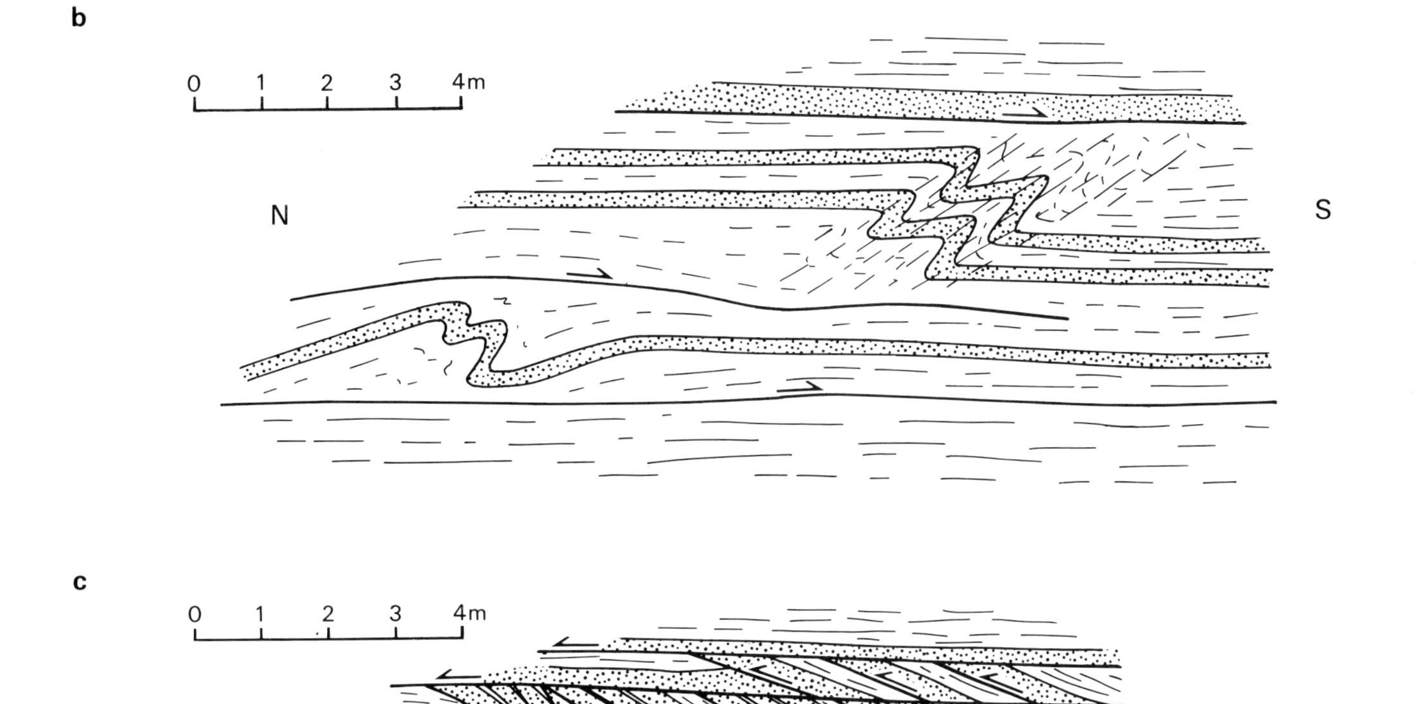

c

0 1 2 3 4m

S

N

Figure 11 Field sketches of structures in the Falmouth district

(a) Pendower Beach [9000 3814]. Tight reclined F_1 folds crossed obliquely by F_4 folds, in chert interbedded with dark siliceous slate of the Pendower Formation.
(b) Western Cove [6567 4522]. Southward vergent thrusts and associated D_3 folds and cleavage in the Porthtowan Formation.
(c) Western Cove [6552 4513]. Southward vergent D_3 duplexes in the Porthtowan Formation

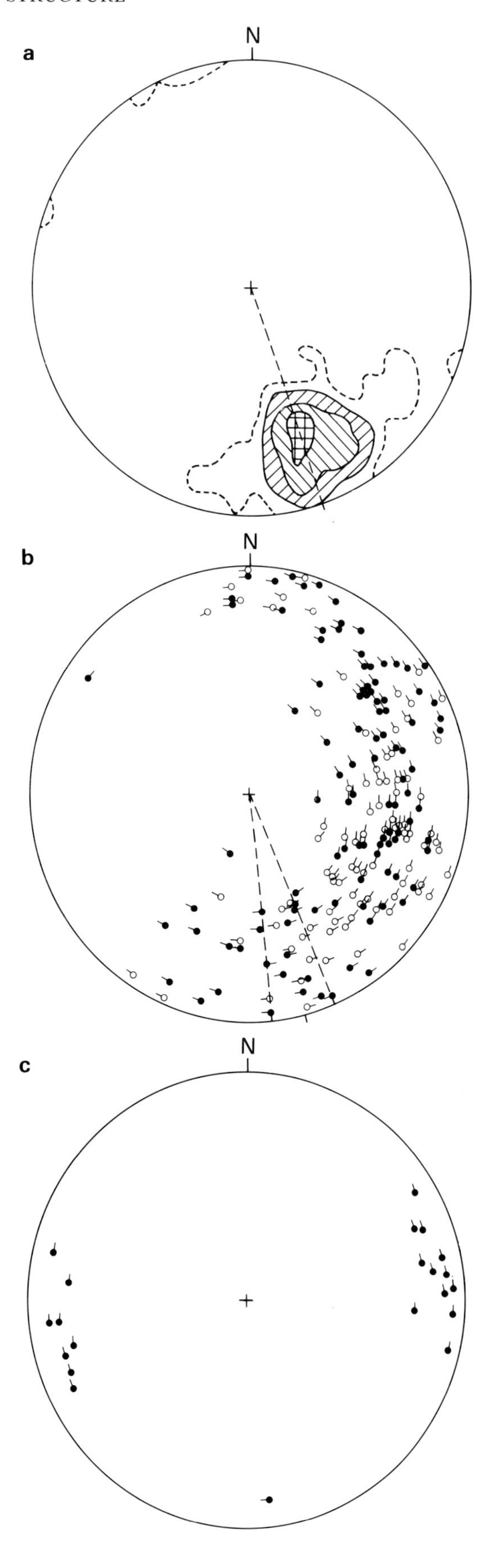

Figure 12 Stereographic projections of D_1 structures

(a) Contoured equal area projection of poles to 46 grain lineations; contours at 5%, 10%, 15%
(b) Poles of 196 F_1 fold axes (solid circles) and bedding/cleavage lineations (open circles) along the south coast. Ticks show facing.
(c) Poles of 21 F_1 fold axes along the north coast

west, and those with an east–west trend facing north, form distinct subdomains in the region (Rattey and Sanderson, 1984). There is also no evidence that sheets of rock, each with its own group of folds of similar trend, behave independently. A distribution pattern consistent with the variation of trend, facing and vergence directions and profile geometry, is represented in the three-dimensional surface diagram Figure 14. This is a tergiversate fold (Dearman, 1969) or sheath fold as redescribed by Quinquis et al., (1978). The curvilinear structure illustrated is monoclinic if the hinge sections of opposing vergence are of equal length. The plane of symmetry is perpendicular to S_1 and contains the south-south-east–north-north-west azimuth. This direction corresponds with the grain lineation (Figure 12a), indicating the key role of extension within the principal plane of flattening in the formation of the structures. In the exten-

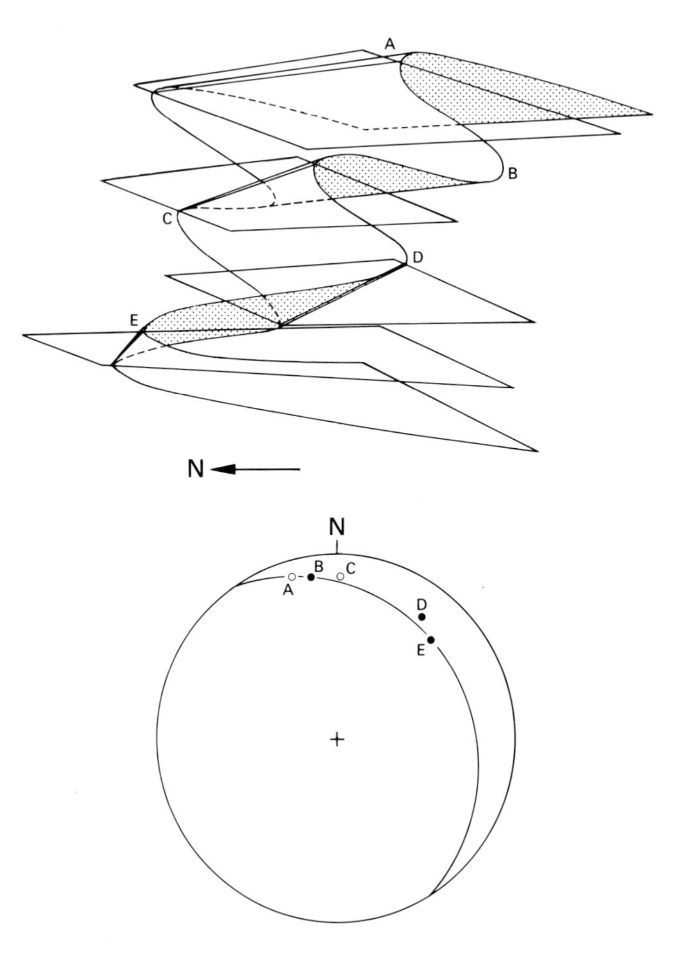

Figure 13 Surface diagram showing the relationship of folds of different trend near Pendower Hotel [892 379], and a stereographic projection of the poles of the fold axes

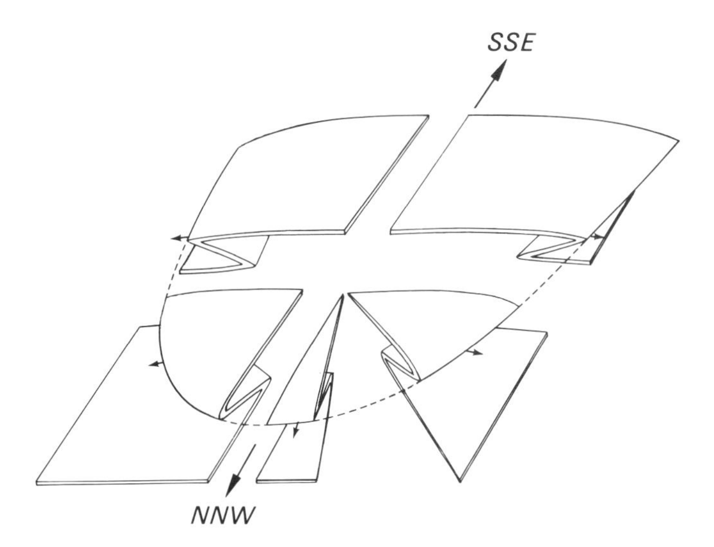

Figure 14 Dissected three-dimensional surface diagram showing the relationship of folds of different trend and facing direction

sion azimuth all profiles and profile components face and verge north-north-west and relative transport of overlying strata to the north-north-west is indicated.

Simple shear mechanisms have been deduced (Rhodes and Gayer, 1977) and experimentally demonstrated (Cobbold and Quinquis, 1980) to be responsible for the formation of sheath folds. Recent work in natural systems, where the strain is not very high, suggests that sheath folds form in a progressive regime, involving both simple shear and flattening (Holdsworth and Roberts, 1984), which appears to be apposite for the south coast. It has been recognised that sheath folds are commonly developed in thrust nappe terrains (e.g. Rhodes and Gayer, 1977).

By way of contrast, the fold axes along the north coast define a strong maximum on the stereographic plot (Figure 12c). These folds are perpendicular to the transport direction and developed under a regime of less sustained shear and flattening. This appears to be consistent with their location in the parautochthon well down in the footwall of the Carrick Thrust. The rocks here would have undergone less deformation than rocks deep in the Carrick Nappe and near to the tectonic source.

Regional metamorphism

The rotation and mechanical disruption of grains, pressure solution and recrystallisation, particularly of micaceous and siliceous minerals in the slates and fine matrix in sandstones, constitutes the regional metamorphism. There has been some subsequent recrystallisation of the rock body during the development of post-D_1 fabrics in the tectonic aureole to the granite batholith, if Turner's (1968) hypothesis is valid, and during thermal metamorphism (see section three).

White mica crystallinity studies in the Falmouth area by Merriman (1982, peronal communication 1986) show that most pelites belong to the epizone, indicative of the greenschist facies of low-grade metamorphism. However micas in pelites of the Portscatho Formation close in the hanging wall close to the Carrick Thrust have a lower degree of ordering, spanning the anchizone–epizone boundary.

In the Mylor Slate Formation and in Gramscatho Group rocks to the north of the district, regional metamorphism has resulted in assemblages of quartz-albite-muscovite-chlorite, a subfacies of greenschist metamorphism (Phillips, 1964). Basic intrusions within the Mylor Slate Formation have had primary assemblages converted to albite-chlorite-epidote and actinolite (Floyd, 1983; Floyd and Al-Samman, 1980) corresponding to low-grade greenschist facies.

The presence of detrital plagioclase in the Pendower Formation sandstone, containing pumpellyite and sericite aligned parallel to S_1, indicated to Barnes and Andrews (1981) a syn-D_1 pumpellyite-actinolite metamorphism similar to that in the greenstones in east Roseland. They concluded that a temperature of about 300°C and a pressure of at least 3 kilobars were operative during the regional metamorphism. This would be consistent with nappe transport from depths up to the 12 km proposed by Holder and Leveridge (1986a).

The K-Ar dates of slate in the Portscatho Formation obtained by Dodson and Rex (1971) cluster around 350 Ma, but using revised methods of calculation these should be nearer 355 Ma (Hawkes, 1981). This corresponds to the Devonian/Carboniferous boundary (Snelling, 1985) and the time of uplift and emergence of the Carrick Nappe.

MAIN PHASES OF POST-D_1 DEFORMATION

Second phase deformation (D_2)

Structures of this phase are present at numerous localities along the south coast and in adjacent areas (e.g. Rosemullion Head, Dearman and others, 1980). On the north coast modifications to F_1 folds are attributed to D_2 but the identification of D_2 structures there is largely based on occurrences along the coast just outside the sheet area.

S_2 cleavage: S_2 is essentially a crenulation cleavage, which characteristically displays a variety of fabrics (Rickard, 1961). In dominantly pelitic rocks it varies from a gentle microflexuring of early fabric to a transposition pseudoslaty fabric (Plate 8c). This is observed as a spatial variation accommodated through several tens of metres of the rock body [8740 3315] or a few centimetres. In semipelites dislocation is more prominent, and mica growth in and adjacent to fractures is evident at an early stage. The S_2 fabric in psammites varies from pressure solution lamina to closely spaced shear planes with mica felts and intervening microlithons showing gentle sigmoidal curving of remnant S_1. Only rarely does S_2 become the dominant fabric in psammites.

Along Carrick Roads a more complex relationship between S_1 and S_2 is apparent where predominantly pelitic rocks are brecciated. Deformation of S_1 fabric in psammite fragments indicates folding prior to disruption. Some clasts are D_2 fold mullions with the same trend as intact D_2 folds elsewhere. The main microtextural feature of these rocks is the closeness of S_1 and S_2 with pseudoslaty S_2 cleavage and subparallel S_1 fabric co-existing in the same rock [E 59876].

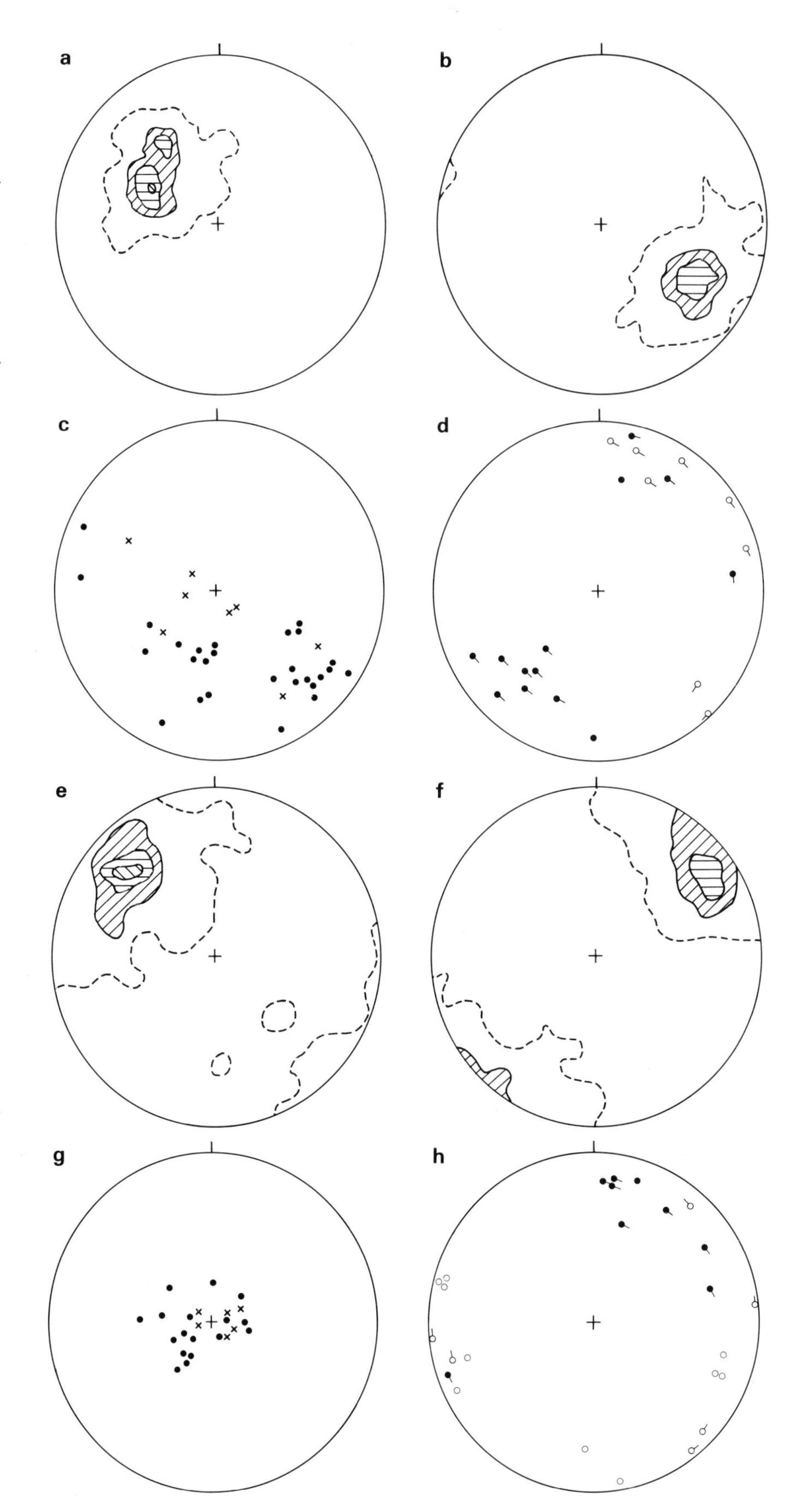

Figure 15 Stereographic projections of $D_2 - D_5$ structures

(a) 78 poles to S_2; contours at 10%, 15%, 20%
(b) 88 poles of F_2 axes; contours at 10%, 15%
(c) Poles to S_3; dots for south coast, crosses for north coast
(d) Poles of F_3 axes; solid circles for south coast, open circles for north coast. Ticks show vergence
(e) 175 poles to S_4; contours at 5%, 10%, 12.5%
(f) 159 poles of F_4 axes; contours at 5%, 10%
(g) Poles to S_5; dots for south coast, crosses for north coast
(h) Poles of F_5 axes; solid circles for south coast, open circles for north coast. Ticks show vergence

Poles to S_2 cleavage are distributed in the north-west quadrant of the equal area projection circle (Figure 15a). A split in the contoured higher concentrations of poles reflects rotation by the late open folds plunging east-south-east. The modal north-east strike and moderate south-east dip is very similar to S_1, and D_2 structures are more prominent in areas where the S_1 and S_2 maxima correspond (e.g. SW 83 SW).

F_2 folds: F_2 folds are north-verging, asymmetrical, mesoscopic-scale folds in which short limbs of less than one metre predominate. They commonly occur as isolated fold couplets; smaller amplitude folds being close to tight and larger folds open. Their axes define an east-south-east trend maximum (Figure 15b) with most folds being reclined or inclined with a high-pitch angle. Fold couplets with common limbs of less than one metre are present at Carne Beach where they form reclined, close, modified parallel folds with rounded hinge zones (Plate 9a). In contrast, at the northern end of Towan beach [874 332] a zone of D_2 transposition some 100 m wide, representing the common limb of a large-scale, north-north-east-verging couplet, crosses the foreshore. Lower order F_2 folds on his structure are close to tight and plunge gently to moderately between east and south-east.

On the north coast at Porthtowan [689 480] subhorizontal F_1 folds trending east–west have been modified by D_2. S_2 is subparallel to S_1. F_2 folds coaxial with F_1 are developed in the hinges and limbs of the F_1 structures. The folds are close to tight with rounded hinge zones and their effect is to open up the limbs of the tight early structures.

Third phase deformation (D_3)

D_3 structures are highly variable across the area, the variation apparently representing an overall decrease in intensity of deformation in a south-easterly direction.

S_3 cleavage S_3 is sporadically developed in association with F_3 folds in zones a few tens of metres wide on the north coast and in zones less than ten metres wide on the southern coast. It is a crenulation cleavage in pelites and a fracture cleavage in psammites, and it dips north-westwards. The dip is gentle to moderate in the north-west, moderate to steep in the south-east (Figure 15c). The cleavage transposes early fabrics at localities on the north coast (e.g. [6400 4460]) whereas in the Falmouth area it is a weak to strong crenulation fabric [8034 3078].

F_3 folds F_3 folds occur as isolated south-east-verging couplets along the north coast (Figure 15d). They are tight to isoclinal [6300 4353], except where they occur in association with D_5 thrusts. Here, the interlimb angles are larger. Short limb lengths of these folds do not exceed 5 m. This is in contrast with F_3 folds in the Mylor Slates and Portscatho Formation to the south-east (Figure 15d), where they are represented by small-scale, open-to-close, asymmetric buckle folds with short limb lengths up to 0.5 m. They occur in zones of south-verging cascades.

Thrusts D_3 is partly represented in the Porthtowan Formation on the north coast in the cliffs below Carvannel Downs [642 448–647 453] by a subhorizontal thrust zone. The zone locally includes duplexes, small blind imbrication stacks and culmination anticlines, and is associated with D_3 folds and cleavage (Figure 11b,c). Both the thrust and the folds verge south-eastwards. Transport in this direction is also indicated by the geometry of the small ramps, duplexes and the culmination structures.

Similar structures occur at several localities in the Fal and Truro river sections (e.g. [856 408 to 858 405]) where folds verging south-eastwards are associated with minor imbrication and duplexes. These folds are open-to-close structures with gently inclined long limbs and subvertical short limbs involving several tens of metres of strata. The phase relationship of these structures, however, is not constrained and it is possible that they may be later structures than D_3 (see below).

Fourth phase deformation (D_4)

S_4 cleavage The fourth phase cleavage is a crenulation cleavage variably developed within well-defined north-easterly zones. The full developmental sequence, from widely spaced crenulations to cleavage banding is rarely present in pelitic rocks. Commonly, S_4 comprises bands, up to 1 cm in which there is closely spaced dislocation, separated by thicker bands in which crenulation is poorly developed. The thinner bands are characterised by a reorientated early fabric and recrytallised micas aligned parallel or subparallel to dislocations, whereas in the thicker bands the early fabric has an overall sigmoidal configuration. This is similar to the fabric in some psammites where spaced dislocations display reorientation and recrystallisation fabrics. Elswhere S_4 cleavage in psammites is a fracture cleavage. Vein quartz and chlorite are commonly associated with the widely spaced cleavage, being locally deformed or displaced by the dislocations or occupying the cleavage planes.

The S_4 cleavage, in common with both S_1 and S_2, strikes in the north-east quadrant, but its moderate to steep dip is generally steeper than earlier cleavages (Figure 15e). In a few places the cleavage is curviplanar being shallower downwards. Where S_2 is also present the two cleavages may, in some cases, be inseparable on a mesocopic scale.

F_4 folds S_4 is generally an axial plane cleavage in folds although a widely spaced crenulation may be present without mesoscopic folding (as at Carne Beach [907 381]). The folds are open to close and asymmetrical, the degree of asymmetry increasing with the decrease in the angle of intersection of the early fabric and S_4. They pitch gently on cleavage, predominantly north-east (Figure 12f) and verge north-westwards. First-order folds vary in scale from very small structures, which are an extension of the crenulation fabric (Plate 9b), to mesoscopic-scale folds with short limbs 2 m in length.

Fifth phase deformation (D_5)

D_5 structures are not strongly developed, but a locally intense S_5 cleavage is encountered in some boreholes in the mineralised belt. Structures in the eastern half of the district

a

b

Plate 9 Post D_1 structures in the Pendower Formation, Pendower Beach

(a) Reclined F_2 folds in chert, limestone and slaty mudstone. Associated S_2 disrupts the thin limestone bed, centre-left foreground, and is itself cut by steeply inclined S_4 spaced cleavage, bottom left. [9065 3824] viewing ENE (b) D_4 structures, small-scale folds and crenulation cleavage in interlaminated limestone and calcareous slaty mudstone. [9075 3823] viewing NE.

are largely restricted to the Mylor Slates, but to the west they extend into the Porthtowan Formation on the coast.

S_5 cleavage S_5 is a crenulation cleavage, subhorizontal to gently inclined to the south-east (Figure 15g). Nowhere in the district does it approach the transposition fabric seen elsewhere in the region (Turner, 1968). The fabric is commonly emphasised in the metamorphic aureole of the granites by mimetic crystallisation.

F_5 folds F_5 folds are small-scale, open, asymmetrical folds with gently plunging axes and amplitudes commonly less than 0.5 m. Hinge zones are either rounded or angular. Vergence defines two distinct provinces; the Carrick Roads area where it is to the south-east, and the north coast where it is to the north and north-north-east (Figure 15h).

OTHER STRUCTURES

There are numerous minor folds and faults, not directly attributable to the main deformation phases, whose interrelationships are uncertain, but which appear to postdate D_4.

North- to north-east-trending folds

Gentle-to-open, small-scale folds with a north to north-east trend are sporadically developed throughout the area. They are upright to steeply inclined with a gentle plunge and a wavelength generally less than one metre. They can be symmetrical or asymmetrical. Adjacent folds are commonly incongruous. Some folds have an associated widely spaced crenulation cleavage or fracture cleavage striking between north – south and north-east – south-west and dipping steeply either side of vertical. This cleavage is observed to deform S_4 along the Carne Beach section.

East- to east-south-east-trending folds

Strike variation in the eastern half of the area are due largely to open folds that plunge between east and east-south-east. These structures are observable on mesocopic scale along foreshore platform sections in Roseland. The elongate and split maxima of contoured S_1, S_2 and S_4 pole plots of that area (see Figures 10 and 15) indicate the presence of the open structures and their post-D_4 development.

Kink bands and related folds

Kink bands of different orientations and attitudes have been recorded in the area. There are three sets: i) north – south-striking bands which dip gently to moderately westwards, and verge eastwards, ii) east – west-striking conjugate kink bands, in which the axial planes dip moderately and the minor angle of the conjugate set is subvertical, iii) north-east – south-west-striking bands dipping very gently south-east with fold vergence to the south-east.

The last group of kink folds appear to merge into a group of minor folds with rounded hinge zones. Observed in a north-east direction these folds have Z-shaped profiles and axial planes generally inclined parallel to or at a gentler angle than bedding or early cleavage fabrics. Dislocations parallel, or subparallel, to the axial planes are common and, although similar in general appearance to some D_3 structures, they deform steeper fabrics, including S_4.

Faults

Two main categories of faults are present in the area apart from the thrusts associated with D_1 and D_3.

Gently to moderately inclined faults This category includes strike faults that in the eastern part of the district dip to the south-east at angles generally less than 35° and are thus less steep than, or parallel to, bedding. Also included are faults in the north coastal area that dip north-west at angles up to 40°. In the intervening mineral belt low-angle lodes thought to occupy faults of this group dip in opposing directions, e.g. The Great Flat Lode dips south-east at 30° and the Wheal Jane elvan and lode dip north-north-west at about 40°.

Some of these faults are simple, planar dislocations; others have an adjacent zone of disrupted rock, and a few contain a thin clay gouge. Folds adjacent to the faults generally indicate normal displacement. Deformation of S_4 adjacent to such faults (e.g. near Gidleywell [909 382]) indicates that the latest down-dip movement postdated D_4. The faults are commonly associated with zones of F_2 and F_4 structures, which may mean that the faults of this group have a polyphase development. Alternatively it may indicate that dip-slip faults are more apparent where they transect folded rocks. There is no evidence of substantial displacement along these faults, because lithological subfacies on both walls are always similar.

Moderately and steeply inclined faults This is the main group of faults. They are predominantly strike faults that cut and displace the lower angle structures. Thus, in the Roseland and Falmouth area north-east – south-west-trending faults predominate although subsidiary north – south and east – west faults are present (Figure 16a). In the west of the district the north-east – south-west faults dominate, but east – west faults become more prominent northwards (Figure 16b). Inland, particularly south of Truro, north-east – south-west lineaments are interpreted as faults, particularly as the Treworgans Sandstone Member is displaced along the Tresillian River lineament just to the north of the sheet. These faults appear to be major regional structures. The fault passing through St Just in Roseland may be in continuity with east-north-east – west-south-west faults along Castle Beach [819 320] that extend south-westwards across Meneage and form the boundary between the Mylor Slate and Portscatho formations. The throw of these faults appears to be mainly down dip, to the south-east on the south-east side of the peninsula, and to the north-west on the north-west side. In the mineralised zone the main lodes and elvans occupy this family of faults and inclinations and displacements are either to the south-east or north-west. Slickensiding within some of these faults (e.g. [8600 3189])) indicates that at least some late movements were lateral.

A subgroup of the category are a set of north-north-west – south-south-east faults in the mineralised belt, com-

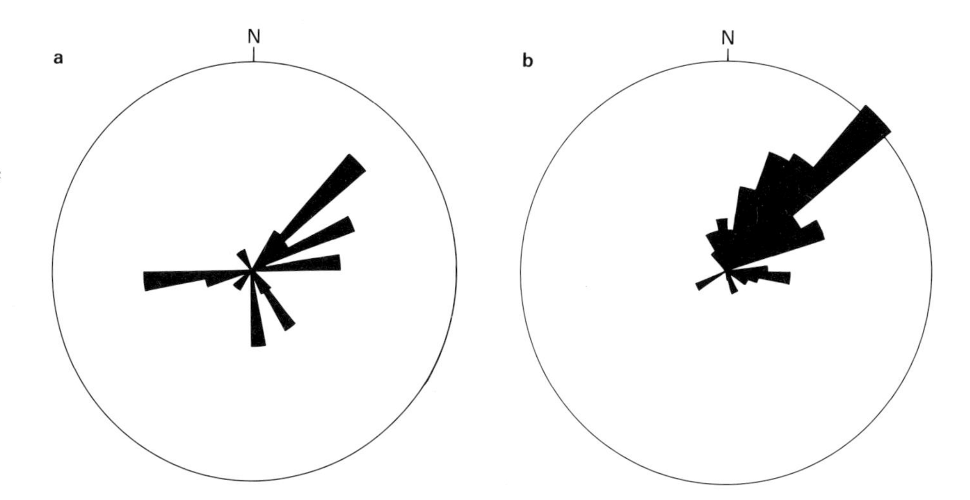

Figure 16 Rose diagrams of measured late faults

(a) 39 faults along the north coast; circle limit 10 faults
(b) 114 faults along the south coast; circle limit 20 faults

monly bearing a lower temperature ore mineral assemblage than the main lodes that they consistently cross-cut. They are the site of the 'crosscourse' lodes such as the 'Great Crosscourse' between Porthcadjack Cove and Bolenowe.

As the north-east – south-west strike faults are occupied by elvans and mineral lodes, they predate the elvans. The change of hade over the granite, suggests that they may postdate the granite intrusion. Their orientation and attitude are consistent with formation in a north-west – south-east extension phase.

Deformation kinematics

Sheath folds are characteristic of highly sheared thrust terrains (Cobbold and Quinquis, 1980). In Cornwall the orientation of the F_1 sheath folds indicates north-north-west-directed overthrusting, which Leveridge et al. (1984) have associated with the emplacement and stacking of the major south Cornish nappes. Rattey and Sanderson (1982), despite their suggestion that D_1 folds are cylindrical, have also concluded that they were associated with northward overthrusting.

S_2 generally has a similar trend to S_1, but a slightly steeper attitude, which suggests a strain regime related to D_1. Certainly, S_2 crenulation fabric represents some shortening perpendicular to cleavage. The presence of a secondary 'shear band' foliation close to the main fabric has been attributed by White et al. (1980) to a strain softening processes, such as continued crystallisation, that produces mechanical weaknesses and new sites for shears. The F_2 folds, however, are cylindrical showing a consistent east-south-east trend and north-north-eastward vergence. As D_2 structures are not regionally penetrative, and major sheath folds are not developed, the F_2 folds indicate rotation in a regime with a shear component to the north-east. In thrust terrains variations in bedding attitude are commonly caused by culmination anticlines (Boyer and Elliot, 1982; Butler, 1982) formed by the movement of thrust sheets over ramps. Rather than D_1 and D_2 being perpendicular, it is possible that continued movement during D_2 locally and temporarily produced obliquity between bedding and the D_2 shear plane. Interference above and in the vicinity of the north-west – south-east sidewall ramp of the Carrick Nappe could have produced a north-east shear element.

Third phase folds and cleavage are closely associated with south-east-verging thrusts. Reversals in thrust transport direction are associated in many thrust belts by the formation of 'triangle zones' (Jones, 1982). These zones form where thrusts terminate and the continued nappe displacement leads to the formation of thrusts with reversed displacement. In south Cornwall these south-verging thrusts and the accompanying D_3 folds in the parautochthon (Mylor Slates) may therefore relate to the formation of triangle zones during continued north-north-west transport of the major thrust nappes following sticking on thrusts beneath the parautochthon. Related folds in the Porthtowan Formation of the north coast of south Cornwall are upper mesoscopic structures verging south and are the structures Smith (1965) designated F_1.

Rattey and Sanderson (1982) indicated that moderate to steep north-west-verging D_4 folds, and the shallowly inclined D_2 folds are parts of the same deformation phase (their D_2) forming discrete shear zones. They regarded these zones as subhorizontal beneath the Lizard Complex steepening northwards, and associated with north-west overthrusting of the Lizard Nappe. However, the localisation of D_4 folds in the zones in which D_2 folding is most intense argues against this hypothesis and suggests that both phases may be related to the presence of underlying thrust ramps. D_4 deformation probably reflects bending of the rocks of the thrust nappe as they were transported over footwall ramps during continued north-north-west overthrusting.

D_5 deformation has been interpreted by Turner (1968) and Rattey (1979) as representing doming and flattening over the south-west England batholith prior to intrusion of the separate cupolas. In support of this theory those authors reported that there is a vergence culmination in F_5 over the batholith in south Cornwall, geographically linked to the granite outcrops, from north vergence in the north to south vergence in the south. This survey has not confirmed that there is a consistent link between F_5 and granite outcrops, there being no intense cleavage zone around the Carmenellis Granite. It is likely that D_5 is related to shear associated with the late Carboniferous backslip on major thrusts in the region (Day, 1985) producing folds during uplift coincident with the early stages of emplacement of the granite.

FIVE

Mineralisation

The intensity of hydrothermal metalliferous mineralisation in the belt of country between Camborne and St Day is unequalled in Great Britain. On completion of the first geological map of Devon and Cornwall in 1839, it was apparent to De la Beche that tin and copper ores occurred in the vicinity of the granite masses whereas ores of lead, antimony, manganese and iron were located at a greater distance from the granites. It was the distribution of ore minerals in this and neighbouring districts that led Dewey (1925) et al. to devise the now 'classic' theories of mineral zonation, which played an important part in the evolution of thought on metallogenesis.

DISTRIBUTION

In the Falmouth district the main tin-copper mineralisation is within a west-south-west – east-north-east belt passing tangentially to the north of the main outcrop of the Carnmenellis Granite (Figure 17), including the smaller granite masses of Carn Brea and Carn Marth and numerous elvan dykes. The dykes appear to occupy a similar set of fractures to the lodes. A north-north-west – south-south-east-trending subsurface granite ridge coincides with an extension of the main mineralised belt between Carn Marth and St Agnes. A substantial area of tin-copper mineralisation occurs in the vicinity of Wheal Vor [625 303] in the extreme south-west of the district. Within the Carnmenellis Granite, tin-copper mineralisation extends sporadically between Wendron [679 311] and Halabezack [703 347]. Lower temperature lead-zinc mineralisation occurs in various crosscourses throughout the district and also in west-south-west – east-north-east lodes in the area between Kerriack Cove [677 470] and Porthtowan, and south-west of Falmouth.

A single north – south-trending lead-silver lode occurs 1 km east of Shortlanesend [808 477] near the northern margin of the district.

TYPES OF MINERALISATION

The main types of metalliferous mineralisation are lodes, replacement ore bodies, disseminated mineralisation, greisen-bordered bodies, quartz-feldspar veins of pegmatitic affinity and metasomatic deposits.

Lodes

Mineralisation is commonly developed along fractures or other planes of weakness as an infilling or as partial replacement of the wallrock by metallic ores. These structures, known as lodes, most commonly originated as faults parallel

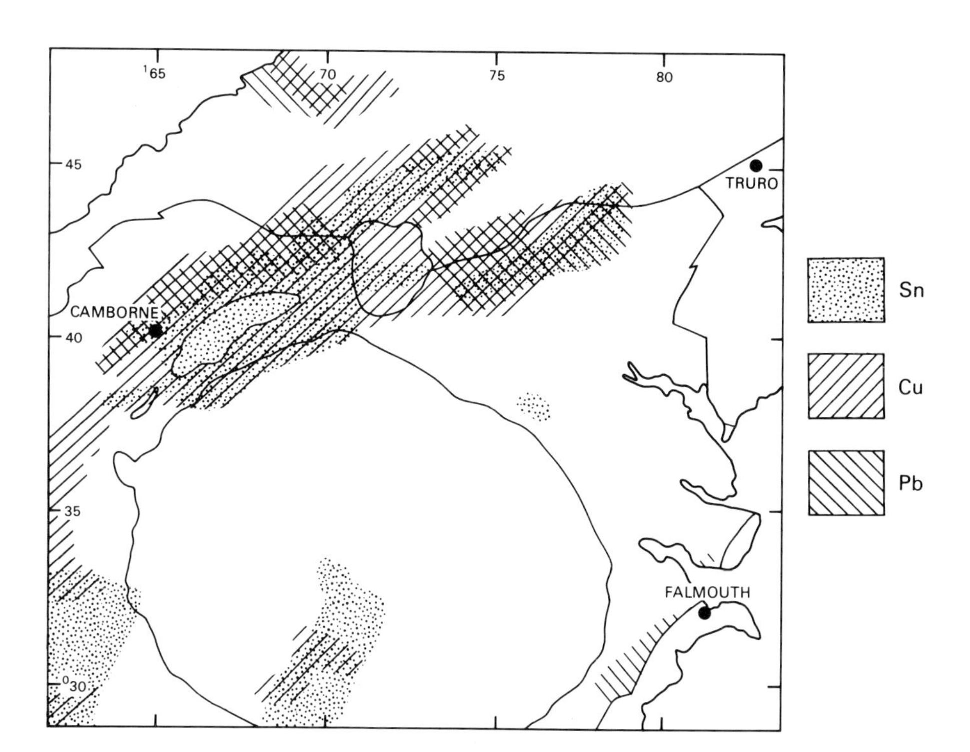

Figure 17 Distribution of mineralisation in the Falmouth district

to the longitudinal axis of the batholith, but some formed along joints, shears, cleavage, bedding or the interface between sedimentary and igneous rocks. Lodes vary in thickness from a few centimetres to about 5 m, but those that have been exploited average just under 1 m wide. The Great Flat Lode, at nearly 6 km in length, is one of the longest mineralised structures known, but the majority are less than 1 km long.

Most lode structures have an east-north-east – west-south-west trend and dip steeply between 70° and the vertical; among the uncommon, gently inclined lodes are the Great Flat Lode at Wheal Grenville, which dips 30° south-east, and the B Lode of Wheal Jane, which dips 40° north-north-west.

Some lodes have long and complex histories involving repeated fissuring, brecciation and resealing by mineral growth. Lodes formed by multiple deposition are characterised by various mineral layers parallel to the vein walls.

The west-south-west – east-north-east lodes are cut and in some cases displaced by faults, known as crosscourses, which trend between north-west – south-east and north – south. Crosscourses are usually filled with either a clay-gouge or low-temperature quartz or a mixture of the two. Ore minerals are generally those deposited at low temperature. Some of these structures probably predate some of the lodes as they have a considerable effect on the richness of the ore minerals contained therein. Wheal Vor [623 301] is one of the best examples of a mine where crosscourses influence mineral concentration within a lode. The crosscourse veins are extensive; the Great Crosscourse, for example, extends from Porthcadjack Cove [641 447] some 8 km south-south-west to near Bolenowe [673 380].

The main primary ore-minerals that make up the higher temperature lodes of the district (e.g. South Crofty [664 410]) are cassiterite, SnO_2, wolframite, $(FeMn) WO_4$, and arsenopyrite, FeAsS. Lower temperature ores include chalcopyrite, $CuFeS_2$, sphalerite, ZnS, and galena PbS (e.g. Porthtowan area). Gangue minerals in the high-temperature deposits, include quartz, which occurs in veins of all types, and feldspar, mica, tourmaline, chlorite and haematite. Fluorspar occurs in both high- and low-temperature deposits, whereas baryte, chalcedony, dolomite and calcite are low-temperature gangue minerals.

Cornish metalliferous deposits have long been interpreted as displaying primary thermal zoning. The zonation is related to the temperature gradient that existed between the granite and the land surface. Minerals that crystallise at higher temperatures, such as cassiterite, wolfram and arsenopyrite, were deposited in lodes in the margins of the granite plutons or in the adjacent county rocks, whereas lower-temperature minerals for example sulphides of copper, zinc, lead and antimony are present in lodes farther away from the granite. Dewey (1925) collated a wealth of information from nineteenth-century Cornish mines and constructed thicknesses for individual mineral zones. Dines (1934, 1956), however, aware that the mineral deposits did not fit neatly into a pattern, proposed that mineralisation was located about emanative centres where pressure/temperature conditions and fissures were well suited for the passage of fluids and the tranpsort and deposition of minerals. Other factors now regarded as important in controlling mineral deposition are the reaction of the hydrothermal fluids with the wall-rock and the composition of these fluids, particularly their salinity. The concept that mineral zones are sharply defined and contain limited suites of ore and gangue minerals is now known to be an oversimplification. Repeated pulses of hydrothermal activity under varying conditions allow a wide range of high- and low-temperature minerals to be deposited in successive stages in the same vein.

Supergene enrichment of lodes occurs by the leaching of metals, particularly copper, above the water table, and their redeposition at a lower level. The zone of leaching is capped by a gossan consisting of gangue minerals and manganese and iron oxides. Below this, but above the water table, the red and black oxides of copper, cuprite and melaconite Cu_2O, CuO, form in the zone of 'oxidised enrichment'. Where copper-rich solutions percolate below the water table the copper sulphides, bornite, Cu_5FeS_4, and chalcocite, Cu_2S, are precipitated by reaction with the original sulphides in the absence of oxygen producing a zone of 'secondary sulphide enrichment'. In the upper parts of some copper lodes of the Camborne – Redruth area, enriched copper ore, such as chalcocite, was the main ore mineral, (e.g. Cooks Kitchen Mine [665 407], Carn Brea Mine [680 410]). Chalcocite has also been named 'redruthite' because of its common occurrence in that area.

The main lodes at South Crofty Mine [666 412] trend west-south-west – east-north-east and have a dip greater than 50°. At depth they contain the high-temperature minerals wolframite, cassiterite, arsenopyrite and chalcopyrite; gangue minerals include chlorite, quartz, tourmaline and hematite. The lodes are composite and contain low-temperature minerals, such as chalcopyrite, galena and sphalerite, in their upper parts. Mineral zonation can be observed locally in mines near South Crofty, which began as copper mines before working tin at depth.

At Wheal Pendarves [647 384] the complexities of the lodes are well illustrated. Tryphena Lode dips 75° south-east and consists of seven veins, four of which have been mined. Harriet Lode consists of ten steeply dipping narrow veins which coalesce at depth to form a 1.5 m wide structure dipping 50° south-east.

At Wheal Jane [772 426] the B lode occupies a shear zone 3 – 10 m wide beneath an elvan which dips about 40° north-west. Steeper lodes dipping at about 70° intersect the B lode and were probably the source of the mineralising fluids in the flatter structure. The lodes of Wheal Jane show a complex depositional history of several phases of hydrothermal activity. The earliest phase resulted in the tourmalinisation of the shear zone and the deposition of quartz and cassiterite. Extensive chloritisation and the introduction or remobilisation of quartz and cassiterite preceded a phase of pyrite, sphalerite and chalcopyrite veining of the earlier ore. A final phase of sulphide deposition occurred in vuggy quartz veins, which crosscut the earlier structures, and commonly carried small amounts of galena.

The Great Flat Lode trends west-south-west – east-north-east from south of Camborne to south of Redruth dipping at about 30°S. It has been worked in many mines, notably Wheal Grenville, West Frances, the Basset group and Wheal

Uny. The lode consists of a central leader, 0.05 – 0.60 m wide, commonly brecciated. This is bordered either above, below or on both sides by 1.20 – 4.60 m of tourmaline-cassiterite-chlorite-quartz rock beyond which the tourmalinised country rock has no tin values. Tin mineralisation is sporadic and occurs in steeply pitching ore shoots which have confusingly been referred to as 'pipes'. The Great Flat Lode is cut by the steeper lodes and is invariably downthrown to the north-west.

Replacement ore bodies

Replacement ore bodies, in the form of cylindrical pipes and horizontal sheets, have been termed 'carbonas' and 'floors' in Cornwall, though the term carbona has been used indiscriminately to describe any large or rich body of ore not obviously in the form of a lode. Replacement usually occurs near a lode or vein and is commonly accompanied by tourmalinisation and kaolinisation. Examples of carbonas from the district include those at Wheal Basset [692 400], Poldice Mine [740 427] Balmynheer [702 346] and South Wendron [301 703].

Disseminated mineralisation

Where permeability is high because of extensive fracturing mineralising fluids may penetrate large volumes of rock to form a 'stockwork' deposit.

At Wheal Music [704 471] the sandstone and slate of the Porthtowan Formation have been microfractured and soaked in copper-bearing fluids. Thin veinlets of chalcopyrite, altered to malachite near surface, have been worked 'opencast' in a 50 m-deep excavation known as the 'navvy pit'.

A well-developed fracture system infilled with cassiterite in an elvan dyke was recorded at South Crofty Mine (Taylor, 1963), and there is a similar elvan at Wheal Vor. This lode, 0.6 m wide in the surrounding slate, has been worked over a width of 6 m within the elvan.

Other examples of this type of mineralisation have been recorded at Pedn an Drea [702 420], Wheal Vyvian [732 293] and Wheal Busy [740 445].

Greisen-bordered bodies

Greisenisation is an alteration process commonly developed at vein margins by the passage of fluorine-rich hydrothermal fluids. The mineralogy of the altered vein typically consists of quartz and white mica with topaz, fluorite and a variety of ore minerals. There are a few examples of greisen in the district. They are identified mostly in old mine workings and are poorly documented, though Hosking (1969) recorded a swarm of greisen-bordered veins over the granite cusp at Carn Brea. Dump material from mines around the northern periphery of the granite suggests that greisens occur near the contact and are more common than hitherto recorded.

At East Wheal Lovell [706 320] there are tin-bearing so-called pipes (Foster, 1878), the largest extending over a depth of nearly 130 m. It is apparently centred on a narrow (1 – 2 cm) vein of ferruginous quartz and clay surrounded by greisenised granite. The mineralogy of the ore from the pipe is similar to the early greisen-bordered veins (Hosking, 1969). These pipes have also been described as 'carbonas' (Dines, 1956) along with similar bodies at Basset and Grylls Mine [690 327].

Quartz-feldspar veins of pegmatitic affinity

Some of the earliest vein deposits consist of wolframite and arsenopyrite-bearing quartz-potassium feldspar veins on the northern flank of the Carn Brea granite.

The 3ABC swarm of veins extends from the deeper levels of New Cook's Kitchen to the upper levels of Robinson's section of South Crofty Mine [667 412]. The east – west-trending vein swarm has a strike length of 670 m, a vertical extent of 330 m, a maximum thickness of about 60 m and a dip of 20 – 30° south-south-east. Individual veins vary in thickness from 0.08 – 0.3 m and are everywhere cut by the main phase lodes of chlorite-cassiterite. These veins have been called pegmatites, but they contain too little feldspar and do not have the right texture.

The Complex Lode swarm occurs in Robinson's section of South Crofty Mine and is sometimes referred to as the Roskear Complex Lode. The steeply dipping vein swarm has an east-north-east – west-south-west-trending strike length of approximately 300 m, a vertical extent of 137 m and a maximum thickness of about 60 m. These structures are also cut by later cassiterite-bearing veins.

It is suggested by Beer and Ball (in press) that similar veins occur in the vicinity of Rogers Lode in East Pool Mine [674 418]. Hosking (1969) believed that the quartz-feldspar veins may have developed above a granite cusp emplaced within the main granite.

Metasomatic deposits

Metasomatic or skarn-type deposits do not constitute a major part of the mineralisation of south-west England, but two examples of note occur within the district.

Stanniferous lodes with unusual mineral assemblages occur within greenstone sills at Magdalen Mine [765 376]. The lodes are up to 0.45 m thick, have a north-west strike and dip 20° north-east. The ore consists of magnetite, hornblende, cassiterite, quartz and chlorite with minor amounts of arsenopyrite and pyrite and traces of bismuthinite and scheelite (Hosking, 1969).

Specimens from the dumps of Trevoole Mine [638 371] indicate a similar mode of occurrence. Dodecahedral pink garnets up to 0.03 m occur with chalcopyrite, pyrite and sphalerite in altered greenstone. Further evidence of metasomatic alteration of the greenstones between Pendarves and Tuckingmill is gained from garnet-diopside-epidote veins in specimens on mine dumps. No other metalliferous mineral deposits are linked with this phase of mineralisation.

AGE OF MINERALISATION

It is generally agreed that the emplacement of the granite was followed by the intrusion of the elvan dykes, which in most cases preceded the deposition of the lode minerals. The relationship of a pegmatite vein cutting an elvan in a borehole at Carwynnen suggests a possible closeness in age of granite and elvan which has yet to be confirmed by isotopic methods. Darbyshire and Shepherd (1985) proposed 269 ± 4 Ma for the age of the main phase of mineralisation. At least three subsequent phases of mineralisation are suggested to have occurred at 225–215 Ma, 170–160 Ma and 75 Ma by Jackson et al. (1982).

ORE GENESIS

Sorby (1858), in his classic work on fluid inclusions in mineral crystals, deduced that metalliferous veins are deposited from saline aqueous fluids at more or less elevated temperatures. Dines (1956) and previous workers considered that the mineralising fluids were derived directly from a granitic magma. Later workers (Sheppard, 1977; Jackson et al., 1982; Alderton and Rankin, 1983; and Shepherd et al., 1985) with the benefit of stable isotope analysis techniques and microchemical and microthermometric analysis of fluid inclusions, have deduced that meteoric and formation fluids play an important part in mineralisation. There is support for this interpretation in the small amount of data available for vein systems in the district (Scrivener, 1986).

An early phase of mineralisation in the working mines of the district, such as the Harriet system of Wheal Pendarves, is cassiterite-tourmaline-quartz veining. Fluid inclusion studies (Scrivener, 1986) show that the veins were deposited from fluids of very variable salinity (2–32 wt % NaCl equivalent) and minimum (uncorrected) trapping temperatures of 250 to 375°C. There is evidence that the highest salinity values are associated with the earliest wallrock tourmalinisation and may indicate the influence of magmatic fluids. The quartz-cassiterite-tourmaline intergrowths, however, were deposited from fluids of lower salinity (2–10 wt % NaCl equivalent) with a large meteoric water component. There is no evidence of boiling in these systems.

Wolframite-quartz assemblages from the 3ABC and Complex lodes of South Crofty Mine give minimum trapping temperatures in the range 275 to 325°C and low salinities (2–10 wt % NaCl equivalent). High CO_2 contents in a proportion of these inclusions suggest that wolframite deposition is linked to the presence of that gas.

Assemblages of cassiterite-chlorite-quartz with sulphides comprise later stages of mineralisation in the Tryphena system of Wheal Pendarves, but are also in the normal lodes of South Crofty Mine and predominate in the lodes at Wheal Jane. These ores are deposited from multiple stages of low salinity (2–10 wt % NaCl equivalent) fluids (Plate 10) with minimum trapping temperatures in the range from 130 to 350°C. The spread of trapping temperatures and the low salinities for this type of mineralisation is in accord with measurements made by Bull (1982) for cassiterite – base metal sulphide ores in east Cornwall and, again, suggests the strong influence of meteoric water.

Both of the preceeding groups of veins have fluid inclusions that are essentially sodium chloride-water filled. In contrast, the crosscourses bear inclusions in which calcium chloride is also important and may comprise up to 50 wt % of the dissolved salt. Measurements for crosscourse quartz from South Crofty Mine demonstrate salinities in the range 8–20 wt % salt calculated as 50% NaCl + 50% $CaCl_2$, and minimum trapping temperatures from 132 to 180°C. Low eutectic melting observed in inclusions from quartz–cassiterite intergrowths at South Crofty Mine indicate that the crosscourse fluids with calcium chloride in solution may have mixed with sodium chloride brines at an early stage. The presence of calcium chloride fluids has been considered by Shepherd and Scrivener (1987) to indicate the large-scale

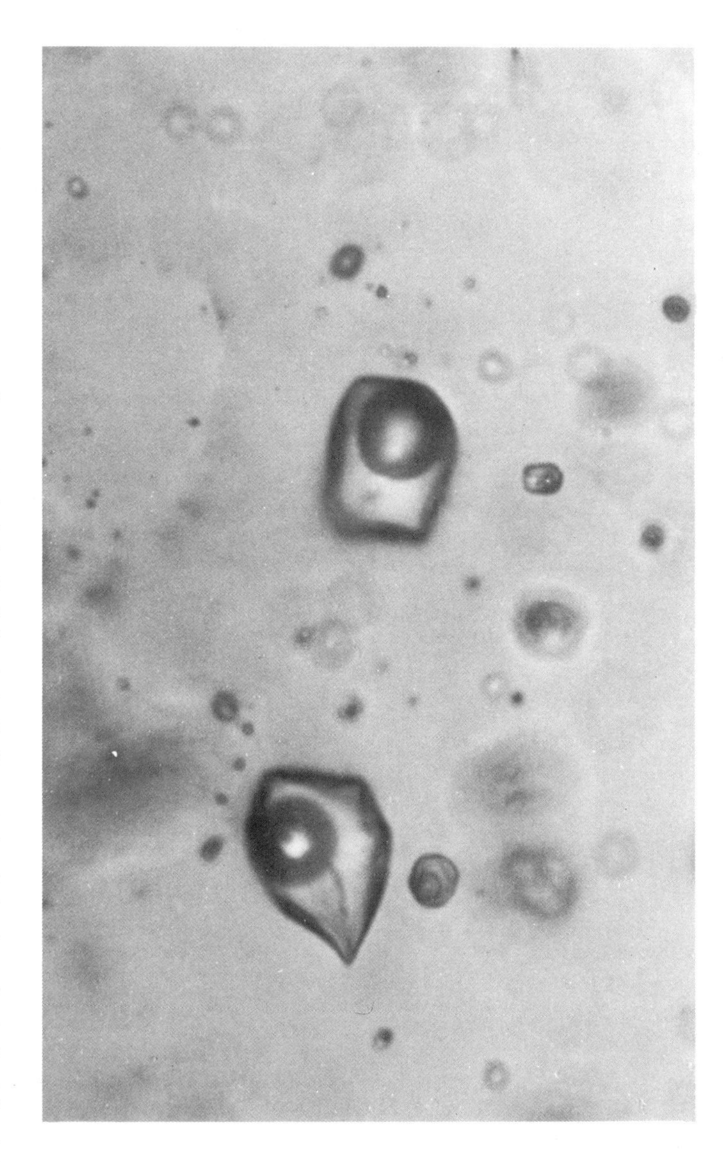

Plate 10 Liquid-vapour inclusions in fragments of quartz from a cassiterite-bearing vein at South Crofty Mine.

The inclusions contain water vapour and a dilute aqueous fluid containing c.4 weight percent sodium chloride equivalents. On heating the vapour bubbles become smaller and disappear between 300° and 350°C. This 'temperature of homogenisation' indicates a minimum trapping temperature for the original hydrothermal fluid. ×900.

involvement of basinal brines during main stage mineralisation in south-west England (Figure 18).

SOURCE OF MINERALISATION

Although the processes of transport and deposition are relatively well understood, the source of the metals which form the hydrothermal deposits of south-west England remains the subject of speculation. There is general agreement that there is a relationship between the mineralisation and the Cornubian batholith, with three possible sources for the metals: a) direct magmatic derivation; b) scavenging by hydrothermal fluids from metal-enriched Palaeozoic country rocks; c) by assimilation into the granite of quantities of the same metal-enriched Palaeozoic rocks.

Mitchell (1974), Floyd et al. (1983), Darbyshire and Shepherd (1985) and Goode and Merriman (1987) all support a derivation of the batholith by crustal melting. The concentration of metals in such a granite melt, derived from an inhomogenous basement, can be accomplished by magmatic differentiation and fractional crystallisation. On crystallisation, the metals reside within the granite, for example tin within biotite. Circulating hydrothermal fluids would subsequently remove and concentrate the tin prior to its deposition in lodes.

The feasibility of Devonian sedimentary rocks being a major source of tin and tungsten appears unlikely. Beer and Ball (1986) concluded that the background levels of both metals are not exceptional, tin being lower than, and tungsten only twice the world average for pelites. However, a source for the copper, lead and zinc in the Devonian country rocks is much more likely. Both the 'Gramscatho' greywackes and the Mylor Slate Formation have higher than average concentrations of lead and zinc, but low copper (Edwards, 1976; Floyd and Leveridge, 1987) whereas metabasalts have a high Cu content, averaging 85 ppm (Floyd, 1968).

A volcanogenic source for copper, lead and zinc must also be considered likely now that the widespread extent of the volcanicity, spanning the peninsula during the Middle and Upper Devonian, is known. Certainly an association has been established in south Devon (Leake et al., 1985) and the geographical association is apparent in south Cornwall. Such an association is characteristic of the underthrust plate environment in continent to continent collisions (Garson and Mitchell, 1977).

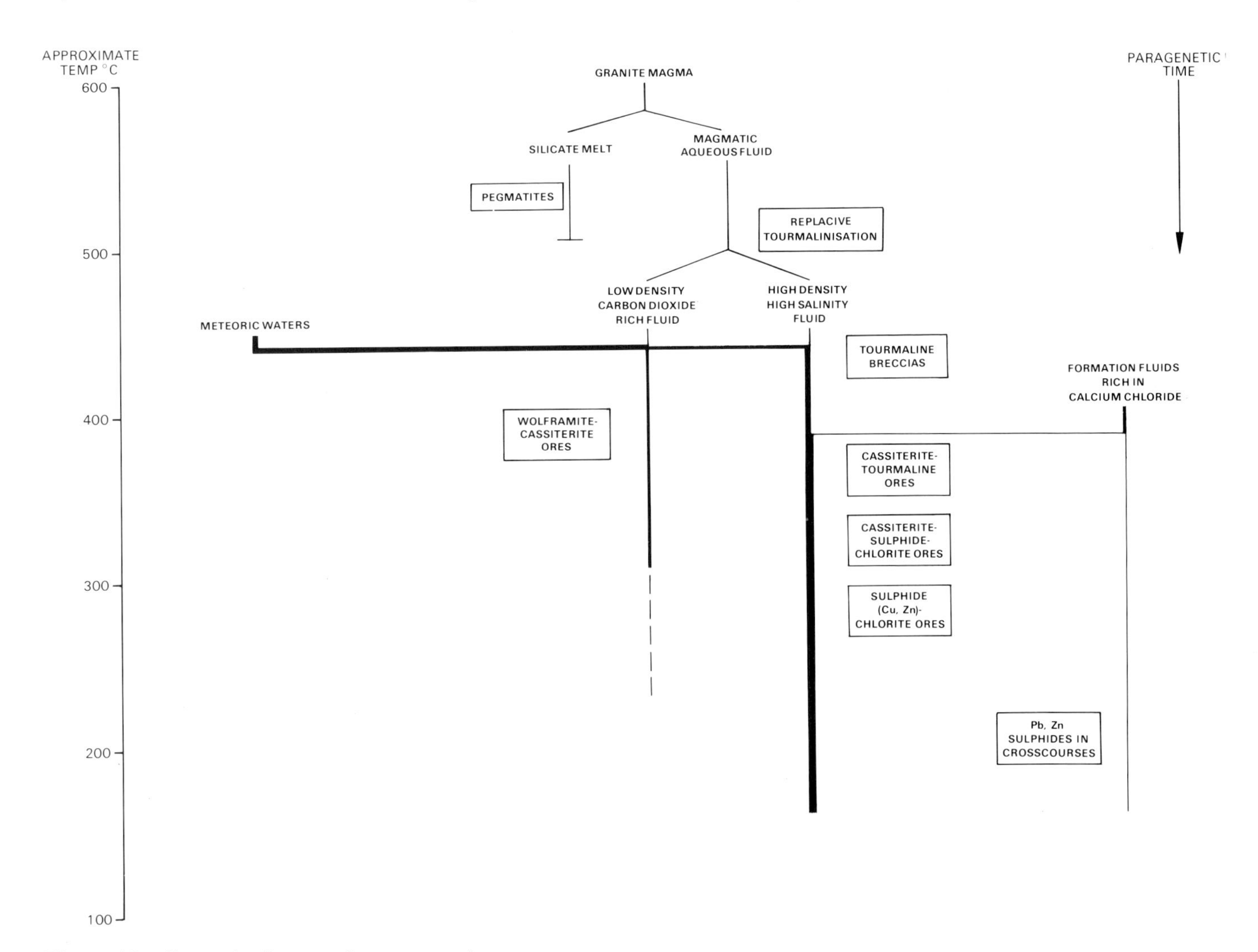

Figure 18 Synoptic diagram for ore genesis

SIX

Tertiary and Quaternary deposits

Apart from minor gravels of probable Tertiary age at Polcrebo and Penhallow, superficial deposits of the district are Pleistocene and Recent in age.

Except over the granites, the solid geology has a limited influence on the topography of the area. Landform is largely a composite of terrace-like features in the solid facing the nearby coast, inlet, valley or inland basin. They occur at intervals of a few metres in height, those at lower levels being stronger and generally less degraded. Their development is thus best observed in the lower areas of the eastern half of the sheet. They tend to follow contours except where two or more amalgamate to form a more pronounced, steeper face. They have been interpreted as relict cliff lines marking the Tertiary and Pleistocene marine retreat from the land area of south-west England (cf. Balchin, 1964; Kidson, 1977).

Recent data show that the maximum fall in sea-level occurred during the late Pleistocene (see Evans and Hughes, 1984). Late Pliocene sea level, as indicated by the St Erth Beds (Mitchell, 1973) is now at c.45 m above OD in south Cornwall, whereas near the shelf edge in the South-Western Approaches that same level is depressed to 110 m below OD (Pantin and Evans, 1984). These authors suggested that this offshore subsidence and uplift of the land area might be related to activity in the asthenosphere. Superimposed on the general uplift of the land during the Pleistocene were major fluctuations of sea-level, with falls related to the removal of water into the ice-caps during glaciations and rises due to its release during the warmer interglacial periods.

TERTIARY

There are no rocks of proved Tertiary age in the district, but two gravel deposits, one a significant spread about Polcrebo [654 329] and the other a small patch at Lower Penhallow Farm [8887 3898], are most probably attributable to the Tertiary.

The Polcrebo Gravels (**Polc**), which is up to 2.4 m thick (Hill and MacAlister, 1906), rests on a plateau feature at 150 m above OD on the Carnmenellis Granite. It is composed of clasts from 10 mm to 1 m in diameter. They are subrounded to rounded and crudely graded. The largest boulders at the base rest on soft yellow clay that may be the product of in-situ weathering of the underlying granite. The boulders comprise not only the vein quartz recorded by Hill and MacAlister (1906) but also mineralised metamorphosed and tourmalinised Mylor Slate Formation sedimentary rocks. At the present time such rocks do not crop out above the level of the gravel. Tyack (1875) interpreted the deposit as alluvial, but Hill and MacAlister (1906) suggested that it represents a shore-line deposit. Those authors assigned a Pliocene age to the gravel, but recognition that sands, gravels and clays at St Agnes, near the northern margin of the Falmouth sheet, are of late Oligocene age (Atkinson et al., 1975) carries implications for the age of the Polcrebo deposits. The fluvial deposits at St Agnes rest on a regional 130 m raised platform tha must be older than the Pliocene (Reid, 1890) or Pleistocene (Weller, 1961) as formerly proposed. It suggests that the Polcrebo deposit at a higher level may be at least of Oligocene age.

The 130 m erosional surface, well developed about the Land's End Granite (Goode and Taylor, 1988), is much less apparent in the Falmouth district although cliff features do occur near Praze-an-Beeble [645 355] and to the north of Carn Brea at this height. More evident is a planation surface between 80 and 90 m above OD along the north coastal region and within the Fal ria drianage basin. An exceptional 30 m of superficial deposits has been encountered in boreholes near Pendarves [650 377] apparently preserved against an erosional feature at c.85 m above OD. It includes head, and minor sand and gravel and may in part represent a residual Tertiary deposit.

A small area of gravel-rich soil at Penhallow is also probably a residual Tertiary deposit on the basis that it occurs at a similar height to the late Pliocene St Erth Beds Formation in the Penzance district (Goode and Taylor, 1988). The gravel is present between 45 and 50 m above OD on a gently rounded col separating two head basins and areas of southerly and northerly drainage. Pebbles and cobbles include locally derived sandstone and vein quartz and exotic quartzite and mica schist. The pebbly sands at St Erth have been interpreted as a beach deposit by Mitchell (1973).

QUATERNARY

Pleistocene

Because the late Pliocene St Erth Beds, deposited around the 40 m level, are said by Mitchell (1973) to represents a phase of marine transgression, it is uncertain that the relict cliff lines mapped below this level are attributable solely to early Pleistocene still-stands. They occur at c.35 m, 20–25 m, 10–15 m and 2.5–5 m heights, corresponding closely with the raised shore platforms assigned to the early Pleistocene elsewhere along the south coast of England (Mottershead, 1977).

Pleistocene deposits rest on the lowest of these raised shore platforms (2.5–5 m) and there are numerous sections through them along the coastal cliffs of Gerrans Bay, Falmouth Bay and Carrick Roads (Plate 11a). From bottom to top the general succession in these sequences is raised-beach deposits, scree, blown sand, head and loess. Manganese-cemented shingle adhering to small remnants of a raised shore platform (c.3 m) in the shelter of the valley at Portreath and in Kerriac Cove are all remnants of these deposits that have survived the active marine erosion along the high, steep, north coastal section.

Raised-beach deposits A coarse clastic deposit, long recognised as a raised-beach deposit (De la Beche, 1839; see also James, 1974), generally forms the lowest part of the sequence on the raised shore platforms. It attains a maximum observed thickness of 3 m at the foot of the fossil cliff at Pendower [8991 3816]. The deposit comprises coarse sandstone, pebbly sandstone and sandy conglomerate with boulders locally present near the base. It is commonly cemented by iron and manganese oxides and at Pendower it forms the roof to caves produced by marine erosion of the underlying inclined siliceous slates and cherts of the Pendower Formation. The beach deposits are crudely bedded, and dip very gently seawards. Cross-stratification is present sporadically. Component pebbles and cobbles are locally derived with vein quartz predominant. No fossils have yet been retrieved from the deposit.

The raised beach of Gerrans Bay has been asigned to the early Devensian or the Ipswichian stage interglacial period by James (1981b), and indeed the sequence of beach deposits, blown sand, soliflucted head would seem to be consistent with marine retreat and onset of cold conditions. There is abundant evidence of submarine cliff features off south-west England marking sea-level still-stands during eustatic marine retreats associated with glacial episodes in the Pleistocene. A particularly well marked feature at 45 m below OD is assigned to the Devensian (see Kidson, 1977) and it would therefore seem likely that the raised beach is a deposit of the phase of maximum rise of sea-level within the preceding Ipswichian interglacial. There is also a possibility that beach deposits, present on more than one of the raised platforms of the south coast of England (Mottershead, 1977), are stranded from the (?) early Pleistocene when the platform were cut.

Scree Generally the raised-beach deposits are succeeded by head but in the Pendower Beach section [8995 3816] a cohesive deposit of local angular platy rock fragments with minor sandy clay matrix is banked against the fossil cliff. It extends over and dies out on the top of the raised beach. Alignment of rock fragments parallel to the upper surface is locally deformed indicating flow at the base of the deposit. Its form is that of a scree fan but it may represent an early deposit of head soliflucted over the cliff onto the raised beach.

Blown sand Resting against the upper surface of the scree and directly on cemented raised beach along the Pendower section is up to 2 m of unconsolidated coarse- to fine-grained sand. It contains a few dispersed small local rock fragments and quartz pebbles, and has an incipient fine bedding. It appears to be blown sand adjacent to the fossil cliff.

Head Head deposits succeed the blown sand in the Pendower section, but elsewhere along the coast head overlies raised beach deposits or rests directly on the raised platform. Sections up to 13 m thick are present at Rosevine [8815 3636] and Rosteague [8775 3375] on the south coast and 4 m of head forms a low cliff on the north coast at Porthtowan [6941 4816]. This head extends up into the main drowned valleys of the district such as the Fal, Truro and Tresillian rivers. It is a heterogeneous deposit largely comprising sandy clay with quartz pebbles and small angular local rock fragments with dispersed blocks. In the coastal sections the basal parts tend to be blocky or contain layers of coarse fragments and in several places a layer with a high concentration of comminuted slate is present towards the top of the sequence (cf. James, 1981a). A crude, thin to massive bedding is generally defined by the proportion of coarser clasts. The bedded is very gently inclined seawards, and platy fragments often display an imbricate structure indicative of flow in the direction of dip. A regional stratigraphy is not apparent, but a hiatus in deposition is suggested at Porthcurnick [8788 3603] where iron pan is present below the upper 1.5 m of head. Involution and ice-wedging in the upper 2 m of the head are particularly well marked in beds of slate fragments at Towan Beach (James, 1981a) and along Pendower Beach.

Cryoturbation in the upper part of the head is interpreted as resulting from freeze-thaw periglacial conditions above permafrost during the Devensian glacial period. Formation of the deposit itself has also been ascribed to the cold phase of the Devensian (James, 1981a). However, the predominance of fine granular material over coarse fragments, preservation of bedding, imbrication of platy fragments and the presence of quartz pebbles in the head indicate active solifluction, with weathering as well as frost shattering, and is more consistent with the less harsh conditions of the onset phase of periglacial conditions.

Loess The disturbed zone of the head is locally overlain by up to 1.5 m of silt with scattered small rock fragments and pebbles. This has been regarded as a loess, having an aeolian origin, but James (1981b) suggested that it has undergone fluvial redistribution. A similar deposit also overlies Recent blown sand at Porthcurnick.

Loess is present on the head in many parts of south Cornwall and has been dated using thermoluminescence methods (Wintle, 1981) as late Devensian.

Holocene (Recent)

Recovery of sea-level following the Devensian glaciation led to submergence and infilling of valleys cut during high run-off phases, particularly at the beginning and end of periglacial considitions. *Buried deposits* in Restronguet Creek and contributary valleys were heavily worked in the past for stream tin (see Hill and McAlister, 1906). The placer tin rested upon solid beneath 15–20 m of varied alluvial deposits (Henwood, 1832). An abundant mixed flora, some in growth positions, and a fauna, including shells and human remains, indicate a paralic environment with subaerial fluvial and estuarine deposition and progressive submergence. There are records of submerged forests at Falmouth, Porthcurnick and Pendower being exposed during exceptional low tide and submerged forest was seen at Portreath during construction of the harbour (see Hill and MacAlister, 1906). Radiocarbon dating of wood from a submerged forest at a similar level in Mounts Bay to the south-west of the present area, has yielded an age of 4278 ± 50 B.P. (Goode and Taylor, 1988).

The inland areas of mapped *head* largely occupy broad gentle depressions at the higher reaches of the main valleys.

Although the depressions are possibly former sites of estuarine or lacustrine deposits associated with higher sea-levels, such deposits do not appear to have survived. In the granite areas the head comprises pale sandy clay containing decaying granite fragments. On sedimentary country rocks the head is largely yellow-beige clay, although some is blue-grey containing numerous ghost slate fragments. Small fragments of lignitic wood are commonly present. Fine sandstone and slate clasts are usually dispersed, but locally define crude bedding. The deposits are probably hill-wash, although the action of small streams in the depressions is marked by lenses and bands of quartz gravel within them. Exotic boulders and blocks within head near Treworthal [8769 3863] are possibly remanié erratics from the early Pleistocene or from the Penhallow gravel.

Over much of the area there is a thin head cover or regolith of local rock fragments in a yellow-beige clay matrix under clay soil. With a decrease of clay downwards the fragmental deposit merges into the solid below. Commonly, the upper few tens of centimetres of solid are deformed due to hill creep. In excavated sections (e.g. [7085 4637]) this thin head blanket is seen to pass beneath the hillwash head, suggesting that it developed mainly in early postglacial times.

The interplay of alluvial processes, both erosional and depositional, in relation to inland head deposits and estuarine deposition along the submerging coast is complex. Differentiation of *alluvial deposits* in the higher levels and lower reaches of the valleys is in some cases arbitrary.

On the Carnmenellis Granite the spreads of alluvium are largely thin covers of redistributed basinal head. The lack of clear slope breaks between head and alluvium indicating accumulation from sheet run-off rather than stream action.

A history of active alluvial deposition by streams draining the granite and mineralised areas is attested by the buried deposits of Restronguet Creek and the Carnon Valley. Deposits at a higher level in the streams in the north-western part of the district have been thoroughly worked for alluvial tin. Between Twelveheads [760 422] and Todpol [742 431] the deposit has been largely removed exposing bedrock.

Remnants of the Pleistocene head that extended up the valleys of the Carrick Roads ria system have been terraced by the downcutting rivers (e.g. [877 471, 824 455]). The spreads of alluvium flanking the main rivers of the area are commonly thin with solid being encountered in the river beds (e.g. [887 422, 870 469]). It is probable, therefore, that alluvial terraces shown at the head of Calenick Creek (e.g. [817 431]) are principally head with a remnant cover of alluvium.

River alluvium is shown on the map backing the creeks at Perranarworthal, Mylor Bridge, Falmouth and Ruanlanihorne. The deposits, like the estuarine alluvium of the creeks, are a complex interdigitation of fluvial and marine detritus (Plate 11b). They are differentiated on the basis of the dominant sedimentary influence now that vegetation is established. Much of the silting up of the creeks is of very recent origin (see Hill and MacAlister, 1906), particularly Restronguet Creek, filled with detritus from the Carnon valley mineral workings, and the Fal valley about Ruanlanihorn, silted up by washings from the St Austell china clay works.

There are accumulations of *blown sand* backing sandy beaches under low modern cliffs (e.g. Towan Beach) and at the mouths of valleys (e.g. Pendower and Porthtowan). At Porthcurnick [8783 3601] some 4 m of blown sand, resting on the Devensian head, are divided by a 0.4 m bed of grey-brown semiconsolidated pebbly sandstone similar in appearance to the Pleistocene beach deposits. The deposits are largely quartz and shell sands with sparce, dispersed whole shells (e.g. Turritellids at Porthcurnick).

The area owes much of its tourist popularity to its sandy beaches, such as those at Pendower, Porthcurnick and Porthtowan. All *beach deposits* are subject to considerable variation with large quantities of finer deposits being added or removed in exceptional tidal and weather conditions. Coarser deposits are of local origin and the sands are largely quartz, with subordinate country rock grains and, in places, a significant bryozoan and shell component.

Plate 11 Quaternary deposits in Roseland

(a) Pleistocene sequence resting unconformably on cherts and siliceous slates of the Pendower Formation at Pendower Beach. Semiconsolidated raised beach deposits are succeeded by unconsolidated *blown sand* (centre) and fine *head* to the top of the cliff. [8998 3816] viewing ENE
(b) Recent *estuarine alluvium* in the Percuil River valley at the junction of Trethem Creek and Polingey Creek. [8618 3431] viewing N. (B E Leveridge)

a

b

SEVEN

Economic, environmental and resource geology

MINING

The long history of exploitation of metalliferous ores in Cornwall probably began in the early Bronze Age when tin was used to harden copper in the manufacture of bronze weapons. Alluvial or stream tin, and lode outcrops were the first sites of mineral workings with underground mining commencing in Cornwall in about the 14th century. Copper production in south-west England reached a peak of some 15 500 tons of metal per annum in about 1860, declining to almost nothing in 1900. Tin production peaked in 1870 at 10 000 tons of metal, falling to 4000 tons in 1900 and to a few hundred tons in 1920 (Dines, 1956). In the Camborne–Redruth mining district, copper production (Figure 19; Table 3) peaked between 1853–1856 and tin between 1891–1893 (Morrison, 1980) (Plate 12). Figures in Table 3 are generalisations because Cornish mine records are usually incomplete. No records of production before 1810 have been given, and where groups of mines have amalgamated it is difficult to assess the total production accurately.

Dines (1956) has summarised the geology, the nature and extent of the workings, and the mineral production of the mines of the district. Historical accounts of mining have been written by De la Beche (1839), Henwood (1843), Collins (1912), Hamilton Jenkin (1962–65) and Morrison (1980).

The following mine details relate to those mines that are current or have been recently active.

South Crofty Mine

The densely concentrated mining properties of the area to the north of Carn Brea have a long and complicated history. The present company operated as South Wheal Crofty from 1854 to 1906 and South Crofty Ltd since. South Crofty took over neighbouring mines including Dolcoath, Tincroft, and East Pool and Agar. Many of the old mines in the vicinity were worked out copper mines that have since proved to contain tin at depth (Table 4). The major interest in the mine was held for many years by the Siamese Tin Syndicate Ltd who acquired the sole interest in South Crofty in 1967. Taylor's section, East Pool, was dewatered during the period 1968–71. In 1971, the Siamese Tin Syndicate Ltd became St Piran Mining Co. Ltd. South Crofty acquired the assets of Camborne Tin Ltd, Camborne Mines Ltd and The Cornish Tin Smelting Co. Ltd in 1973. By this move South Crofty took control of Wheal Pendarves, which was subsequently operated by their subsidiary company, Great Western Ores Ltd. In 1978, further mineral rights were gained by the take-over of Tehidy minerals. In 1982 South Crofty PLC became a subsidiary of Charter Consolidated Ltd, which held a controlling interest of 60 per cent and Rio Tinto-Zinc Ltd with 40 per cent. In 1984, Rio Tinto-Zinc Ltd, through their subsidiary Rio Tinto-Zinc Ltd, acquired Charter Consolidated's 60 per cent interest giving RTZ's subsidiary company, Carnon Consolidated Ltd, control of Wheal Pendarves, South Crofty and Tehidy Minerals Ltd. A decline from the Tuckingwill valley was begun in 1984 aimed eventually at replacing Robinsons Shaft. Permission was sought in 1985 to drill exploratory boreholes on the Great Flat Lode, south-west of Redruth.

Pendarves Mine

Old workings known as Pendarves Consols, Pendarves United and Tryphena occur between Pendarves [646 379] and Beacon [657 393]. Cornish Explorations began re-examining this area in 1962 and a drilling programme commenced in 1963 with limited examination of old workings. Various mining companies financed the venture including Bibis Yukon, Guggenheim Exploration Co. Ltd, Pacific Tin Consolidated Corporation and Tehidy Minerals Ltd. Union

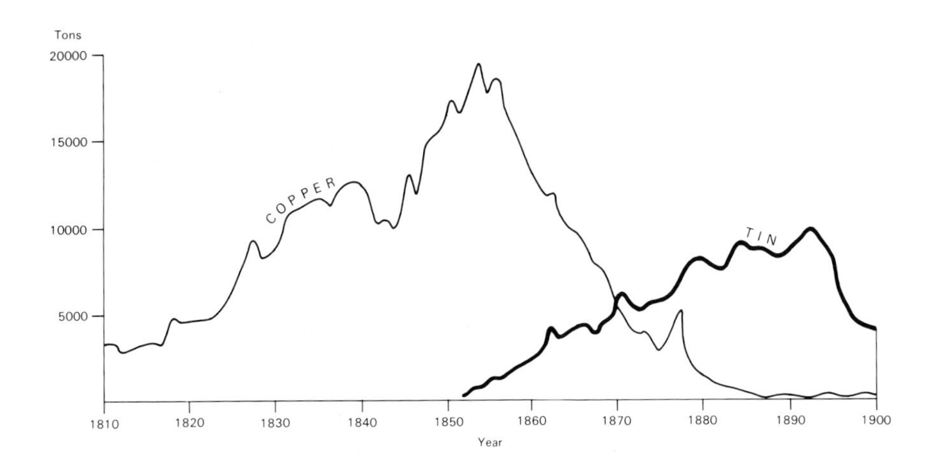

Figure 19 Tin and copper ore production, Camborne–Redruth area, 1810–1900

Table 3 Production figures for copper[1] and tin[2] in the Camborne–Redruth district

Name of mine	Dates of working	Tons of copper ore % metal	Tons of black tin[3]
Bassett Mines	1815–1905	290 000 + (7–8%)	43 000 +
Wheal Buller and Beauchamp	1820–1875	147 100 (6%) 98 700 (14%)	
Carn Brea and Tincroft	1815–1896	360 000 (6–7%)	53 000
Clifford Amalgamated	1861–1870	105 156	
Dolcoath	1815–1905	350 000 (6–7%)	80 000
East Pool and Agar	1836–1919	88 300 (5%) 3022 (13%)	40 461
Wheal Gorland	1815–1853	58 160 (9%)	
Great Consolidated	1815–1857	442 400 (8%)	
Great Wheal Busy	1815–1867	104 700 (5–8%)	
Great Wheal Vor United	1853–1895		15 000 +
Grenville United	1860–1910		14 620
Wheal Jewel	1815–1853	58 160 (9%)	
North Crofty (East Crofty)	1832–1899	100 952 (7%) 9170 (6%)	
North Roskear	1816–1874	167 400 (8%)	
Penstruthal	1825–1879	59 500 (4%)	
Poldice	1815–1849	108 698 (6%)	
South Condurrow	1864–1902		11 430
Wheal Seton	1834–1876	113 050 (6%)	
Tresavean	1815–1885	228 000 (3–9%)	
Treskirby	1815–1927	66 000 (9–9%)	
West Wheal Seton	1848–1890	125 770 (8%)	
United Mines	1815–1861	397 667 (6–7%)	

1 Mines with copper production in excess of 50 000 tons
2 Mines with tin production in excess of 10 000 tons
3 Black tin is a general term for a concentrate of cassiterite

Corporation (UK) Ltd joined the consortium in 1966 to manage the project. The consortium, under the name of Camborne Tin Limited, drilled vertical boreholes on the site of the proposed Pendarves Shaft upon which work was begun in late 1967. The shaft, first to be sunk in Cornwall for 40 years, was completed to 260 m and commissioned in 1969. In 1970 the company managing the mine became Camborne Mines Ltd and started underground exploration by crosscutting north-west and south-east to Harriet and Tryphena lodes respectively. A decision to commence production was made and in 1971 Straus Exploration Incorporated obtained additional finance for this purpose.

Also in 1971, the Cornish Tin Smelting Company Ltd was purchased in order to process the Pendarves ore at their Roscroggan mill. 432 tonnes of ore were produced but operations were suspended in 1972 and the mine was taken over by Great Western ores Ltd, a subsidiary of South Crofty (St Piran Ltd) in 1973. The shaft was deepened to 6-level in 1976 and further exploration of Harriet and Tryphena lodes carried out. By 1973 the shaft had been sunk to 7-level and exploration of Tryphena Lode yielded poor results. A drilling programme was initiated in the same year to examine the Great Flat Lode in the area west of South Tolcarne Mine [656 386], and possible extensions of Dolcoath South lodes. As a result of the international tin crisis of 1985/6 the mine was closed and placed on care and maintenance on 31 July 1986.

Table 4 Tin production of South Crofty Mine

1854 to 1905			5080 tons black tin	
	tonnes ore treated		tonnes black tin	
years	range	mean	range	mean
1906–1945	9306– 75 668	60 250	113– 876	618
1946–1954	28 517– 60 547	41 041	303– 590	487
1955–1971	67 626–111 953	91 048	711–1176	940
1972–1985	145 000–295 743	212 899	1100–1817	1336*

* tonnes tin metal

Wheal Jane

The mineral lodes of the Baldhu area, formerly worked by about a dozen mines which combined in 1906 under the name of Falmouth Consolidated, came under the control of Consolidated Gold Fields Ltd in 1964. Exploratory drilling began in 1966 and shaft rehabilitation and underground exploration followed in 1968. In 1969 a decision was taken to bring the mine, to be known as Wheal Jane, into production in 1971. The mine produced four main metals, Sn, Cu, Sn and Ag (Tables 5 and 6). Annual production of tin peaked in 1973 at 1600 tonnes, but steadily declined to 951 tonnes in 1977 before closure in 1978. Closure was officially brought about by the demise of Mount Wellington Mine, which would increase the already considerable burden of pumping at Wheal Jane. Production figures also show that output from Wheal Jane had declined probably due to hasty exploitation of rich ore and a lack of development of ore reserves.

Plate 12 Derelict 19th century mine buildings in the Carnkie area, near Camborne.

On the left are buildings of the former South Wheal Frances, and centre/right are engine houses of West Wheal Basset. Direction of view W. (A J J Goode).

Table 5 Metal production of Wheal Jane in tonnes of metal

	Year	Sn	Cu	Zn	Ag
	1971	150			
	1972	1260			
Cons.	1973	1600	390	2000	
Goldfields	1974	1480	430	300	2.3
	1975	1276	496	2379	2.6
	1976	1073	465	3340	2.08
	1977	951	335	3257	1.40
	1978	255	109	2047	0.35
	1979				
Carnon	1980	626	238	4332	
Consol.	1981	1499	607	10 855	3.04
	1982	1664	636	10 186	3.04
	1983	1627	652	8879	2.46
	1984	1863	657	7159	2.58
	1985	*	596	5044	1.72
	1986	1822	602	5605	3.0
	1987	1662	750	6522	2.0

* Figure not available

Carnon Consolidated Tin Mines Ltd, a subsidiary of Rio Tinto-Zinc Corporation Ltd, acquired Wheal Jane in 1979 from Consolidated Gold Fields Ltd, and the shaft and underground workings of Mount Wellington Mine from the Receivers of Cornwall Tin and Mining Limited. The mines were refurbished to operate as one unit, using the mill at the Wheal Jane site (Plate 13a). In 1980 production restarted, reaching 1499 tonnes in 1981 and 1863 tonnes in 1984.

Mount Wellington Mine

International Mine Services Ltd leased mineral rights in the Twelveheads [760 423] to Gwennap [738 402] area in 1967.

Table 6 Production of tin and copper ores from former mines of the Wheal Jane area, in tons

	Sn	Cu
Nangiles (+ Wheal Andrew) 1845 – 1908	193	3020 (6%)
W Wheal Jane 1854 – 1889	410	50
Wheal Jane 1847 – 1895	3832	740 (4%)
Wheal Falmouth 1829 – 1833	—	1808 (4.75%)
Wheal Sperries 1829 – 1831	—	3146 (5%)
Falmouth & Sperries 1832 – 1872	—	1150 (3.5%)
Wheal Hope 1824 – 1837	—	5584 (8%)
E Wheal Falmouth 1830	—	47
Falmouth Consolidated, 1905 – 1919	915	—
North Wheal Jane 1861 – 1875	70	—
Wheal Baddern 1848 – 1853	30	—

Diamond drilling commenced in the following year, under the guidance of an associate company, Prado Exploration Ltd. Some 40 boreholes were drilled and a decision to sink a 4.5 m diameter shaft was taken in 1969. A decision to go into production was made in 1974 and 85 tonnes of tin in concentrates were produced in 1976, rising to 420 tonnes in 1977. In April 1978 mining ceased as the grounds that ore reserves and grades were substantially below those predicted.

Wheal Concord

In 1980, Wheal Concord Ltd was formed to examine tin-bearing lodes in the area between Mawla [703 458] and Blackwater [735 460]. The shaft at Wheal Concord (formerly Wheal Briton) [728 460] was rehabilitated and shallow cross-cutting commenced. In 1981, some 8000 tonnes of ore were hoisted. Operations were suspended in 1982 but the mine was not allowed to flood. A new company, Concord Tin Mines Ltd acquired the mine in 1984 and almost immedia

b

Plate 13 Current and recent operations of the minerals industry

(a) The modern headgear of No.2 and Clemow's Shafts and buildings, including the mill (skyline right), at Wheal Jane, Baldhu. [765 430] viewing ENE (b) Alluvium and mine tailings being worked for tin in the Carnon Valley. [7860 4064] viewing NW from viaduct.

ately began underground exploration. The same company commenced the evaluation of leases in the Cligga area [740 535], to the north of the district with the intention of milling the ore from both these projects at a new mill under construction at Gilberts Coombe [690 443].

Training mines

Two mines are maintained solely for the training of students in the techniques of mining. The Royal School of Mines uses Main Lode above adit of Tywarnhayle Mine for mine surveying training. Part of South Condurrow Mine around Williams' Shaft and including workings on Williams' and Kings' lodes was taken over by the School of Metalliferous Mining, Camborne (now Camborne School of Mines) in 1897. It was renamed King Edward Mine at the time of the coronation in 1901 and has been worked periodically to provide ore for mineral dressing training as well as surveying and training of mining practise.

Mine waste and tailings

Small companies specialising in the extraction of tin from dumps and tailings tend to be transient. In recent years there has been activity at Roscroggan [647 418], Gilbert's Coombe [690 443], which has been used since the early 1800s, near Treskillard [668 395] and at Tolgarrick [660 412].

Alluvial tin deposits

Detrital cassiterite was recovered from many of the alluvial deposits, which, in places, have been reworked several times. Little is known in detail of the workings and there are no known production figures. Cassiterite commonly occurred in gravel resting immediately on bedrock 'shelf' below alluvial silt, sand and gravel. In some areas there was more than one tin-bearing horizon but the basal gravel was usually the richest. Detrital tin, liberated solely by natural erosional processes, tends to be pure, unlike metallic ores from alluvial deposits consisting of tailings from mineral dressing operations, which invariably contain metallic sulphides. The principal tracts of alluvium worked for tin are Porkellis Moor [688 324], the Carnon Valley around Bissoe [778 413] and Restronguet Creek [795 388]. Hydraulic Tin, of Bissoe, who worked alluvial tin, mine waste and tailings were taken over in 1978 by Billiton Minerals, a Royal Dutch Shell subsidiary. The mill at Bissoe was modernised and in 1979 Billiton also acquired the mill at Mount Wellington Mine which closed in that year. Preparation to work the Carnon Valley deposits between Bissoe and the Truro–Falmouth road at Devoran continued in 1980 (Plate 13b) and the deposits of Restronguet Creek were prospected before the company ceased operations in 1981.

Concord Tin Mines of Blackwater [728 460] evaluated deposits of alluvial tin from broad areas of south Cornwall in 1985–86.

Marine alluvial tin

The seabed sediments off the northern coast of the district contain cassiterite derived from tailings slimes as well as naturally eroded material. Both the streams at Portreath and Porthtowan flow through major parts of the mining area and their waters have been used in the past for the washing of ores.

Marine Mining (Cornwall), who examined the sediments off the north coast, conducted trial dredging and have produced some ore. It is proposed to preconcentrate the stanniferous sediments on board ship prior to piping the product ashore to be upgraded. Work commenced in 1985 to replace equipment and storage facilities at the former treatment plant at Gwithian [585 422], near the mouth of the Red River.

Recent mineral exploration

In 1963 and 1965 the Union Corporation subsidiary, Camborne Tin Ltd began drilling at the Grenville Mines to the north of Troon, at Wheal Radnor and the Great North Downs Mines and in the Wheal Vor area.

The Barcas Mining Company undertook a diamond drilling programme at Wheal Peevor [707 442] in 1966. Their attempt to dewater the mine ended prematurely when torrential rain flooded it in July 1968.

In 1966–67 the Cornish Tin Smelting Company Ltd commenced a geochemical sampling programme followed by shallow diamond drilling in the Carnkie area. Pumping tests were carried out at Lyle's Shaft, North Wheal Basset.

The Union Corporation subsidiary, Cornish Land Ventures, in the period 1964–67 drilled some 30 boreholes scattered around the western side of the Carnmenellis Granite as far as Marazion in the neighbouring district.

ENVIRONMENTAL GEOLOGY

A legacy of the mining industry is a large area of dereliction and hidden hazards (Figure 20). Attempts made to landscape and plant worked and spoil-covered areas of the mineralised belt (Shipman, 1984) have had only limited success because of toxins produced by arsenic and base metal sulphides in the soil. About 3500 shafts were mapped during the survey. This is an incomplete record and indicates the dimension of the problem that the shafts may present. Open mine shafts in the more frequented areas of the sheet have been capped in recent years. Others have been blocked near surface, but decay of retaining materials will inevitably lead to further collapse. Ground instability is also locally compounded by past mining operations in lodes and stockworks close to surface. Records for many of the larger and more recent mines are available for consultation, but there are none for earlier shallow and potentially more unstable mining excavations.

RESOURCES

Minerals

Production of tin over the last forty years has increased steadily, reaching a peak of 95 000 tonnes of metal in 1985 (Figure 21). In January 1986 South Crofty's ore reserves from more than thirty structures were reported in excess of 3.5 million tonnes at a grade of c.1.5 per cent tin metal.

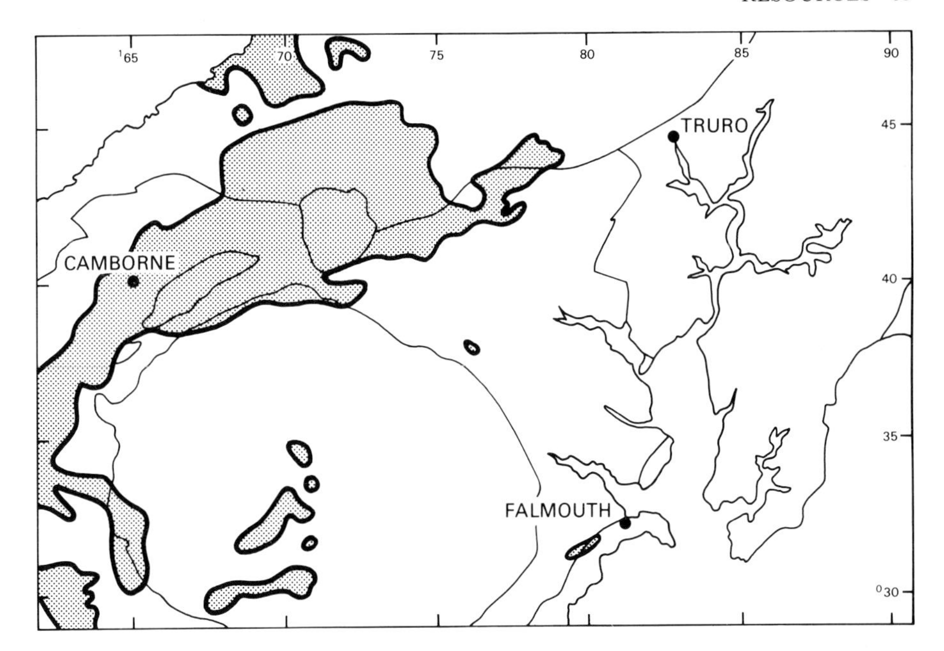

Figure 20 Main areas containing shafts and undermining in the Falmouth district

Although Wheal Pendarves closed in 1986, mainly because the grades of ore were too low, South Crofty has several other interesting prospects in hand including the investigation of parts of the Great Flat Lode. Tin values in the steeper lodes of Wheal Jane improve with depth as the workings approach the eastward extension of the Carn Marth granite. Ore reserves for Wheal Jane in January 1986 were nearly 3 million tonnes at c.1 per cent tin metal, with economic grades of zinc and copper. Similar reserves have been maintained on an annual basis since 1979.

Substantial reserves of tin in alluvial deposits and mine waste have been identified by Wheal Concord, and Marine Mining has proved alluvial tin on the sea floor off the north coast of the district. In the future, interest, however, may lie elsewhere in the district, if, as it is suspected, some copper proves to be volcanogenic in origin.

Quarrying

In the last century, quarrying of the Carnmenellis Granite formed an important part of a thriving Cornish stone industry. The granite, though not superior in quality to the other granite masses, had the advantage of being close to the ports of Falmouth and Penryn. Quarries still working blocks for building purposes or producing sawn slabs for facing are Bosahan, [7295 3031] Carnsew, [759 345] Trenoweth [7587 3388] and Trevone [743 324]. The granite is also worked for aggregate at Carnsew [759 345] and Chywoone [748 348]. A quarry on the south-eastern side of Carn Marth [716 407] was worked in the 1970's and there are current proposals to reopen it.

Small amounts of a low-grade china clay were formerly worked from a quarry at Lower Fergilliock [7835 3315] near Penryn. The kaolin deposit there was the product of the decomposition of G_e granite that forms a minor intrusion on the margin of the Carnmenellis pluton. Hill and MacAlister (1906) also refer to small-scale china clay workings in the granite on Porkellis Moor and St Day, and to the china clays from decomposed elvans around Carharrack, Lanner, Sparry Bottom, Carnon Downs and Wheal Baddon, but no trace of these workings now remains.

Elvan dykes have also been worked for building stone, yielding a wide variety of rock types. Highly altered yellow-buff elvan has been used for buildings in parts of Truro.

In the past, sandstone was dug on a small scale from numerous quarries for the construction of farm buildings and field boundary walls. Some quarries still provide sandstone for repair works.

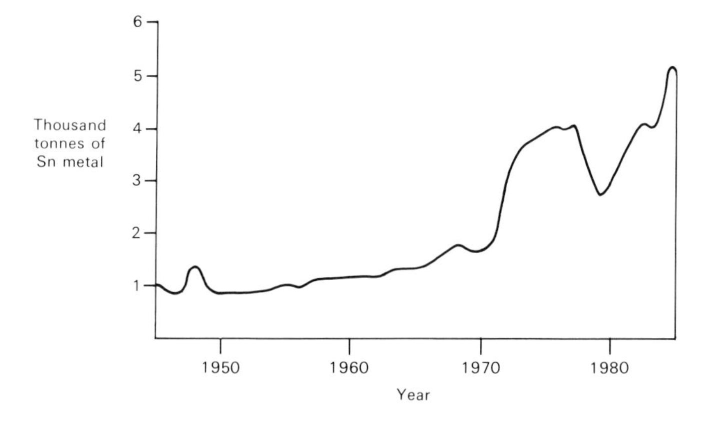

Figure 21 Cornish tin production, 1945–1985

Soils

The soils of the area tend to be thin and reflect closely the underlying geology. Over the sedimentary rocks they are silty-clay or clayey-silt with dispersed rock fragments of local origin. They are generally well drained except in the valleys on clayey head deposits. Their composition is such that they may be seasonally waterlogged or baked, the latter particularly where the soil is thin. Local, thicker, more humic soils in these areas appear to be the product of the past practice of kelping, evidence of it being exotic pebbles restricted to particular fields. In the granite areas the loamy soils are sandy and acid. A humose surface layer is present in the well drained marginal areas but on the higher ground a thin wet peaty cover is common.

Water

The Falmouth district lies within Hydrometric Areas 48 and 49, the water resources being managed by the South West Water Authority. The mean annual rainfall is of the order of 1100 mm, while the mean annual loss to evaporation is between 500 and 550 mm.

The area to the west is drained by small rivers radiating from the high ground of the granite and the east of the district is drained by the river Fal and its tributaries. River flow gauging stations on the rivers Kenwyn at Truro [820 450] and Kennal at Ponsanooth [762 377] show base-flow indices in excess of 0.65, the flows being supported by storage both in the granites and in the country rock. A further significant contribution to base flow is provided by storage in the large number of old mine workings.

A brief mention of groundwater was made by Hill and MacAlister (1906) in the previous survey and conditions of subsurface storage and groundwater flow have been considered by the South West Water Authority (1979) and by Heath (1985). An hydrogeological and hydrochemical map of the Carnmenellis granite and its surrounding country rock has been prepared recently by the British Geological Survey (in press), and this is accompanied by an explanatory report (Smedley et al.).

The bulk of the water put into supply in this district comes from surface sources in the form of the Stithians [718 363], Argal [762 325] and College [767 333] impoundment reservoirs located upon the granite outcrop. The licensed abstraction from these reservoirs amounts to some 13 800 000 cubic metres per annum (m^3/a); in 1987, the actual amount abstracted was of the order of 9 200 000 m^3/a. Underground sources comprise in the main adits and disused mine shafts within the granite and country rock. A total of about 2 170 000 m^3/a is licensed from these sources, of which 2 156 000 m^3/a is for public supply and almost all the remainder for agricultural use; the public supply sources were not used during 1987, being held in reserve against drought demands. There is one borehole source licensed for a substantial take; this consist of two boreholes (together with an adjacent spring) licensed for an annual take of 176 000 m^3 to supply a creamery near Camborne [636 406]. However, there are a number of borehole sources supporting small local demands of up to a few tens of cubic metres per day.

While neither the granites nor the country rock in this district can be termed aquifers, usable groundwater is present both in the weathered zone and in fissures in bedrock. The weathered zone on the granite outcrop is variable in thickness, being absent on the hilltops and up to a few metres in the valleys (Buckley and Cripps, 1989). In the country rock, the weathered zone is unlikely to be more than some 10 or 15 m thick. Beneath the weathered zone, groundwater flow is predominantly along the major joints (Heath, 1985).

The hydrogeological conditions of the granite and the country rock are similar for groundwater at depths less than 100 m. The rock matrices are in general either impermeable or only slightly permeable, and groundwater is contained in, and flows through, joints and faults. Vertical permeability appears to be very much less than the horizontal. Moreover, all the fissure systems may not be interconnected; in the Mount Wellington – Wheal Jane mine complex, dewatering does not seem to affect water levels in adjacent shallow wells. It seem probable that the permeable zones connected with the mine workings extend along the major joint alignments with very limited lateral extension.

The depth of water boreholes in the district does not bear any direct relationship to the depth of the potentiometric surface. It reflects the depth necessary to intersect, at any given location, water-bearing dissures. In the granite, recorded water borehole depths range from 10 to 45 m with an average of about 20 m. In the country rock, depths range from 10 to more than 100 m (although boreholes of more than some 80 m depth were generally unsuccessful) with an average of about 30 m. Borehole diameters are almost all in the range of 100 to 150 mm.

An analysis of the yields of 73 boreholes into the granite shows that they follow a log-normal distribution. The mean yield was calculated to be 37 cubic metres per day (m^3/d), with approximately a 20% probability of the yield being less than 19 m^3/d. A similar analysis of yields of boreholes constructed in the country rock showed a mean of 34 m^3/d and a 20% probability of less than 22 m^3/d. Although the two means are very similar, there is a statistically significant difference between them at the 95 per cent level. No statistically significant differences could be found between yields from the different formations in the country rock.

Shallow groundwaters in both the granite and the country rock have chemical compositions that relate mainly to the bedrock lithology, degree of metalliferous mineralisation, and, to a lesser extent, farming practice. Groundwaters from the country rock generally have a higher total dissolved solids content and major ion concentration than those from the granite. Typical ranges of total dissolved solids for the granite are from 55 to 370 milligrammes per litre (mg/l), and for the country rock from 75 to 525 mg/l. In both rock types, the groundwater is soft. Granite water is usually more acidic (pH usually between 4 and 7) than that of the country rock (pH typically 5 to 8). Where deep-seated groundwater is discharged from mine adits, the acidity is increased. Groundwater in the granite appears to be of the sodium-chloride type, tending towards the calcium-sodium-sulphate-chloride type. The country rock water is of the sodium-chloride type, but tends towards the calcium-bicarbonate type, possibly with distance from the granite. Some typical analyses are shown on Table 7.

Metalliferous mineralisation in the district has given rise to local concentrations of trace metals such as copper and zinc; while the concentrations of these elements are usually less than 0.1 and 0.3 mg/l respectively, copper levels of up to 2 mg/l and zinc up to 16 mg/l have been recorded. Locally, high concentrations of iron (up to 22 mg/l) and manganese (up to 6 mg/l) are commonly associated with the more acidic waters. While little is known about the distribution of arsenic in the groundwaters of the Falmouth district, this element does occur in the mineralised zones and could in theory be found in the groundwater. In the adjacent Penzance sheet

Table 7 Analyses of groundwaters from the granite and the country rock of the Falmouth district

Location	Halabezack Farm [704 354]	Murree [721 413]	Boderluggan [687 314]	Coopers [796 427]	Gweal Mellin [742 282]	Venton Valse [776 507]
Aquifer	Granite	Granite	Granite	Country rock	Country rock	Country rock
Date	19 Aug 88	10 Oct 88	9 Aug 88	3 Oct 88	16 Aug 88	31 Aug 88
Temperature (°C)	15.6	10.8	14.5	12.5	18.5	13.8
pH (pH units)	6.78	5.13	5.70	6.94	6.62	6.32
SEC (μS/cm)	200	210	112	210	120	340
Total dissolved solids	97.2	70.2	83.4	92.5	128.0	197.9
Calcium (Ca)	13.8	5.2	5.7	12.6	9.0	26.4
Magnesium (Mg)	2.2	2.8	5.7	3.3	8.7	12.1
Sodium (Na)	13.4	13.7	18.8	13.9	30.9	20.6
Potassium (K)	2.2	1.4	3.0	2.8	3.8	1.6
Bicarbonate (HCO_3)	17.7	9.2	3.35	7.8	1.7	64.2
Chloride (Cl)	24	22	28	23	41	36
Sulphate (SO_4)	24.7	12.6	12.3	27.6	23.4	25.5
Nitrate (NO_3-N)	<0.8	3.3	6.5	1.5	9.5	11.5
Silica (Si)	1.0	4.1	4.1	2.2	11.4	2.5
Aluminium (Al)	<0.10	0.33	0.31	<0.10	<0.10	<0.10
Strontium (Sr)	0.042	0.025	0.045	0.032	0.053	0.118
Barium (Ba)	<0.002	0.005	0.015	0.009	0.003	<0.002
Lithium (Li)	<0.004	<0.004	0.009	0.010	0.122	<0.004
Baron (B)	<0.02	<0.02	0.03	0.02	0.02	<0.02
Iron (total Fe)	<0.01	0.67	0.05	0.01	0.06	<0.01
Manganese (Mn)	<0.001	0.044	0.035	0.004	0.013	0.014
Copper (Cu)	<0.01	0.01	0.06	<0.01	0.04	<0.01
Zinc (Zn)	<0.01	0.23	0.03	0.18	0.05	0.09
Phosphorus (P)	<0.1	<0.1	0.04	<0.1	0.3	0.1
Total hardness ($CaCO_3$)	43	24	37	45	58	116
Carbonate hardness ($CaCO_3$)	34	13	14	31	22	66

Units in milligrammes per litre unless otherwise stated. Hardness calculated according to Hem (1970, p.224).
Analyses by the British Geological Survey

(351/358) district concentrations of arsenic of 0.01 mg/l As are not uncommon in river and stream waters (Goode and Taylor, 1988).

The use of agricultural fertilisers has significantly affected nitrate levels. These are generally less than 0.8 mg/l (as NO_3-N), but local concentrations of up to 20 mg/l are not unusual.

Geothermal energy

World-wide power generation from geothermal sources is increasing at the rate of 17 per cent per year compared to an annual increase of only 4 per cent from conventional energy sources. Two types of potential geothermal energy sources are presently being exploited: high enthalpy reservoirs in areas of very high heat flow, and low enthalpy reservoirs, where higher than average geothermal heat flows provide deeper level sources of hot rocks or water reservoirs. The latter type of reservoirs. The latter type of reservoir is represented in the United Kingdom where the geothermal heat flow above the south-west England granite batholith, at 120 mW/m^2, is significantly higher than that observed anywhere else in the United Kingdom (Downing and Gray, 1986).

Recent research by the Camborne School of Mines at their site at Rosemanowas Quarry [735 346] in the Carnmenellis Granite has been directed towards solving the engineering problems associated with the extraction of this geothermal heat from the impervious granite by the enhancement of natural fractures in the rock and the injection of recovery of water heated by the hot rock reservoir. Results from the pilot stage of this work have been sufficiently encouraging for the UK Department of Energy to finance a full-scale trial with 6 km-deep wells recovering water at more than 170°C and generating approximately 5 MW of electricity. If this stage of the programme is successful it is calculated (Ledingham, 1986) that the entire south-west England batholith could represent a major new energy resource equivalent to 6000 million tons of coal, or the entire UK national coal reserves.

REFERENCES

Most of the references listed below are held in the Library of the British Geological Survey at Keyworth, Nottingham. Copies of the references can be purchased from the Library subject to the current copyright legislation.

ALDERTON, D H M, PEARCE, J A, and POTTS, P J. 1980. Rare element mobility during granite alteration: evidence from south west England. *Earth and Planetary Science Letters*, Vol. 49, 149–165.

— and RANKIN, A H. 1983. The character and evolution of hydrothermal fluids associated with the kaolinised St. Austell granite, SW England. *Journal of the Geological Society, London*, Vol. 140, 297–309.

AL-RAWI, F R J. 1980. A geophysical study of deep structures in southwest Britain. PhD thesis, Unversity of Wales.

AL-SAMMAN, A H. 1980. Petrology and geochemistry of some volcanic rocks from S. Cornwall. MSc thesis, University of Keele.

AL TURKI, K, and STONE, M. 1978. Petrographic and chemical distinction between the megacrystic members of the Carnmenellis granite, Cornwall. *Proceedings of the Ussher Society*, Vol. 4, 182–189.

ATKINSON, K, BOULTER, M C, FRESHNEY, E C, WALSH, R T, and WILSON, A C. 1975. A revision of the geology of the St. Agnes outlier, Cornwall (Abstract). *Proceedings of the Ussher Society*, Vol. 3, 286–287.

BALCHIN, W G V. 1964. The denudation of chronology of south-west England. 267–281 in *Present views of some aspects of the geology of Cornwall and Devon.* HOSKING, K F G, and SHRIMPTON, G J (editors). (Blackford, Truro: Royal Geological Society of Cornwall.)

BARNES, R P. 1983. The stratigraphy of the sedimentary melange and associated deposits in South Cornwall, England. *Proceedings of the Geologists' Association, London*, Vol. 94, 217–229.

— 1984. Possible Lizard-derived material in the underlying Meneage Formation. *Journal of the Geological Society, London*, Vol. 141, 79–85.

— ANDREWS, J R. 1981. Pumpellyite-actinolite grade regional metamorphism in south Cornwall. *Proceedings of the Ussher Society*, Vol. 5, 139–146.

— — and BADHAM, J P N. 1979. Preliminary investigations of south Cornish melanges. *Proceedings of the Ussher Society*, Vol. 4, 262–268.

BATHER, F A. 1907. Discovery in west Cornwall of a Silurian Crinoid characteristic of Bohemia. *Transactions of the Royal Geological Society of Cornwall*, Vol. 13, 191–197.

BEER, K E, and BALL, T K. In press. Tungsten mineralisation and magmatism in SW England. *Chronique de la Recherche Miniere.*

— — 1986. Tin and tungsten in peltic rocks from SW England and their behaviour in contact zones of granites and in mineralised areas. *Proceedings of the Ussher Society*, Vol. 6, 330–337.

BIRPS and ECORS. 1986. Deep seismic reflection profiling between England, France and Ireland. *Journal of the Geological Society, London*, Vol. 143, 45–52.

BOTT, M H P, DAY, A A, and MASSON-SMITH, D. 1958. The geological interpretation of gravity and magnetic surveys in Devon and Cornwall. *Philosophical Transactions of the Royal Society*, 251A, 161–191.

— HOLDER, A P, LONG, R E, and LUCAS, A L. 1970. Crustal structure beneath the granite of south-west England. 93–102 in *Mechanism of igneous intrusion.* NEWALL, G, and RAST, N (editors). *Geological Journal Special Issue*, Vol. 2.

— and SCOTT, P. 1964. Recent geophysical studies in south-west England. 25–42 in *Present views on some aspects of the geology of Cornwall and Devon.* HOSKING, K F G, and SHRIMPTON, G J (editors). (Blackford, Truro: Royal Geological Society of Cornwall.)

BOUMA, A H. 1962. *Sedimentology of some flysch deposits.* 168pp. (Amsterdam, London, New York: Elsevier.)

BOYER, S E, and ELLIOTT, D. 1982. Thrust systems. *Bulletin of the American Association of Petroleum Geologists*, Vol. 66, 1196–1230.

BRAMMALL, A, and HARWOOD, H F. 1932. The Dartmoor Granites: their genetic relationships. *Quarterly Journal of the Geological Society, London*, Vol. 88, 171–237.

BRITISH GEOLOGICAL SURVEY. In press. Hydrogeological and hydrochemical map of the Carnmenellis Granite. 1:50 000. (Keyworth, Nottingham: British Geological Survey.)

BROOKS, M, DOODY, J J, and AL-RAWI, F R J. 1984. Major crustal reflectors beneath SW England. *Journal of the Geological Society, London*, Vol. 141, 97–103.

BUCKLEY, D K, and CRIPPS, A C. 1989. Geophysical logging of boreholes in the Carnmenellis Granite, Cornwall. *British Geological Survey Technical Report*, WK/89/4.

BULL, B W. 1982. Geology and mineralization of an area around Tavistock, south-west England. PhD thesis, University of Exeter.

BUTLER, R W H. 1982. The terminology of structures in thrust belts. *Journal of Structural Geology*, Vol. 4, 239–246.

CHAPPEL, B W, and WHITE, A J R. 1974. Two contrasting granite types. *Pacific Geology*, Vol. 8, 173–174.

CHAROY, B. 1986. The genesis of the Cornubian Batholith (south-west England): the example of the Carnmenellis Pluton. *Journal of Petrology*, Vol. 27, 571–604.

CHAYES, F. 1955. Modal composition of two facies of the Carnmenellis granite. *Geological Magazine*, Vol. 92, 364–366.

COBBOLD, P R, and QUINQUIS, H. 1980. Development of sheath folds in shear regimes. *Journal of Structural Geology*, Vol. 2, 119–126.

COLLINS, J H. 1879. On the geological structure of the northern part of the Meneage Peninsula. *Transactions of the Royal Geological Society of Cornwall*, Vol. 10, 47–59.

— 1912. Observations on the west of England mining region. *Transactions of the Royal Geological Society of Cornwall*, Vol. 14, 677pp.

— and Collins, H F. 1884. On the geological age of central and west Cornwall. *Journal of the Royal Institute of Cornwall*, Vol. 10, 47–59.

Cowan, D S, and Page, B M. 1975. Recycled Franciscan material in Franciscan melange west of Paso Robles, California. *Bulletin of the Geological Society of America*, Vol. 86, 1089–1095.

Dangerfield, J, and Hawkes, J R. 1981. The Variscan granites of south-west England: additional information. *Proceedings of the Ussher Society*, Vol. 5, 116–120.

— Rundle, C C, Morgan, G E, and Beer, K E. In press. Lamprophyre dykes and associated rocks from south-west England.

Darbyshire, D P F, and Shepherd, T J. 1985. Chronology of granite magmatism and associated mineralisation, SW England. *Journal of the Geological Society, London*, Vol. 142, 1159–1178.

Day, G A. 1985. The Hercynian evolution of the south west British continental margin. *In* International Symposium of Deep Structure of Continental Crust: Results from Reflection Seismic. Barazangi, M, and Brown, L (editors). *Geodynamics*, No. 14.

— and Edwards, J W F. 1983. Variscan thrusting in the basement of the English Channel and SW Approaches. *Proccedings of the Ussher Society*, Vol. 4, 432–436.

Dearman, W R. 1969. Tergiversate folds from south-west England. *Proceedings of the Ussher Society*, Vol. 2, 112–115.

— 1971. A general view of the structure of Cornubia. *Proceedings of the Ussher Society*, Vol. 2, 220–236.

— Leveridge, B E, and Turner, R G. 1969. Structural sequences and the ages of slates and phyllites from south-west England. *Proceedings of the Geological Society, London*, Vol. 1654, 41–45.

— — Rattey, R P, and Sanderson, D J. 1980. Superposed folding at Rosemullion Head, South Cornwall. *Proceedings of the Ussher Society*, Vol. 5, 33–38.

De le Beche, H T. 1839. Report on the geology of Cornwall, Devon and west Somerset. *Memoir of the Geological Survey of Great Britain.*

Dewey, H. 1925. The mineral zones of Cornwall. *Proceedings of the Geological Association*, Vol. 74, 107–135.

Dickinson, W R. 1982. Compositions of sandstones in circum-Pacific subduction complexes and fore-arc basins. *Bulletin of the American Association of Petroleum Geologists*, Vol. 66, 121–137.

— and Suczek, C A. 1979. Plate tectonics and sandstone compositions. *Bulletin of the American Association of Petroleum Geologists*, Vol. 63, 2164–2182.

Dines, H G. 1934. The lateral extent of the ore-shoots in the primary depth zones of Cornwall. *Transactions of the Royal Geological Society of Cornwall*, Vol. 16, 269–296.

— 1956. The metalliferous mining region of south-west England. *Memoir of the Geological Survey of Great Britain.* 2 vols.

Dzulynski, S, and Walton, E K. 1965. *Sedimentary features of flysch and greywackes.* 274pp. (Amsterdam, London, New York: Elsevier.)

Dodson, M H, and Rex, D C. 1971. Potassium-argon ages of slates and phyllites from south-west England. *Quarterly Journal of the Geological Society, London*, Vol. 141, 315–326.

Downing, R A, and Gray, D A (editors). 1986. *Geothermal energy — the potential in the United Kingdom.* 187pp. (HMSO for the British Geological Survey.)

Edwards, R P. 1976. Aspects of trace metal and ore distribution in Cornwall. *Transactions of the Institution of Mining and Metallurgy (Applied Earth Science)*, Vol. 85, 1383–1390.

Evans, C D R, and Hughes, M J. 1984. The Neogene succession of the South Western Approaches. *Journal of the Geological Society, London*, Vol. 141, 315–326.

Exley, C S, and Stone, M. 1964. The granitic rocks of South-West England. 131–184 in *Present views of some aspects of the geology of Cornwall and Devon.* Hosking, K F G, and Shrimpton, G J (editors). (Blackford, Truro: Royal Geological Society of Cornwall.)

— — and Floyd, P A. 1983. Composition and petrogenesis of the Cornubian granite batholith and post-orogenic volcanic rocks in south west England. 153–177 in *The Variscan fold-belt in the British Isles.* Hancock, P L (editor). (Bristol: Adam Hilger.)

Flett, J S. 1933. The geology of the Meneage. *Memoir of the Geological Survey of Great Britain, Summary of Progress for 1932*, part 2, 1–14.

— 1946. Geology of Lizard and Meneage. *Memoir of the Geological Survey of Great Britain.*

Flinn, D. 1965. On the symmetry principle of the deformation ellipsoid. *Geological Magazine*, Vol. 102, 36–45.

Floyd, P A. 1968. Distribution of SiO_2, the alkalis, total Fe and Mg in the granitic plutons of Devon and Cornwall. *Transactions of the Royal Geological Society of Cornwall*, Vol. 20, 21–44.

— 1981. *Geochemical comparison of basaltic rocks in the Lizard ophiolite and the Cornubian troughs* [Abstract]. Tectonic Studies Group meeting on the Lizard Complex. (London: Geological Society of London.)

— 1983. Composition and petrogenesis of the Lizard Complex and pre-orogenic basaltic rocks in southwest England. 130–153 in *The Variscan Belt in the British Isles.* Hancock, P L (editor). (Bristol: Adam Hilger.)

— 1984. Geochemical characteristics and comparison of the basic rocks of the Lizard Complex and the basaltic lavas within the hercynian troughs of SW England. *Journal of the Geological Society, London*, Vol. 141, 61–70.

— and Al-Samman, A M. 1980. Primary and secondary chemical variation exhibited by some west Cornish volcanic rocks. *Proceedings of the Ussher Society*, Vol. 5, 68–75.

— Exley, C S, and Stone, M. 1983. Variscan magmatism in southwest England — discussion and synthesis. 178–185 in *The Variscan fold belt in the British Isles.* Hancock, P L (editor). (Bristol: Adam Hilger.)

— and Lees, G J. 1973. Ti-Zr characterisation of some Cornish pillow lavas. *Proceedings of Ussher Society*, Vol. 3, 489–494.

— and Leveridge, B E. 1986. Lithological and geochemical characteristics of some Gramscatho Group turbidites, south Cornwall [Abstract]. *Proceedings of the Ussher Society*, Vol. 6, 421.

— — 1987. Tectonic environment of the Devonian Gramscatho basin, S. Cornwall: framework mode and geochemical evidence from turbiditic sandstones. *Journal of the Geological Society, London*, Vol. 144, 531–542.

— and Winchester, J A. 1975. Magma type and tectonic setting discrimination using immobile elements. *Earth and Planetary Science Letters*, Vol. 27, 211–218.

Foster, C Le Neve. 1878. On the Great Flat Lode South of Redruth and Camborne, and on some other tin-deposits formed by the alteration of granite. *Quarterly Journal of the Geological Society, London*, Vol. 34, 640–653.

Franke, W, and Walliser, O H. 1983. "Pelagic" carbonates in the Variscan Belt—their sedimentary and tectonic environments. 77–92 in *Intracontinental fold belts.* Martin, H, and Eder, F W (editors). (Berlin: Springer-Verlag.)

Garson, M S, and Mitchell, A H G. 1977. Mineralisation at destructive plate boundaries; a brief review. 81–97 *in* Volcanic processes in ore genesis. *Special Publication of the Geological Society, London*, No. 7.

Ghosh, P K. 1934. The Carnmenellis Granite: Its petrology, metamorphism and tectonics. *Quarterly Journal of the Geological Society, London*, Vol. 90, 240–276.

Goode, A J J. 1973. The mode of intrusions of Cornish elvans. *Report of the Institute of Geological Sciences*, No. 73/7. 8pp.

— and Taylor, R T. 1980. Intrusive and pneumatolytic breccias in south-west England. *Report of the Institute of Geological Sciences*, No. 80/2. 23pp.

— — 1988. Geology of the country around Penzance. *Memoir of the British Geological Survey*, Sheets 351 and 358 (England and Wales).

— and Merriman, R J. 1987. Evidence of crystalline basement west of Land's End granite. *Proceedings of the Geologists' Association*, Vol. 98, 39–43.

Hall, A. 1982. The Pendennis peralkaline minette. *Mineralogical Magazine*, Vol. 45, 257–266.

Hamilton Jenkin, A K. 1962–65. *Mines and miners of Cornwall.* 392pp. (Truro: D B Barton.)

Hawkes, J R. 1967. Rapakivi texture in the Dartmoor granite. *Proceedings of the Ussher Society*, Vol. 1, 270–272.

— 1968. *In* Geology of the country around Okehampton. *Memoir of the Geological Survey of the United Kingdom*, Sheet 324 (England and Wales). Edmonds, E A, Wright, J E, Beer, K E, Hawkes, J R, Williams, M, Freshney, E C, and Fenning, P J.

— 1981. A tectonic 'watershed' of fundamental consequence in the post-Westphalian evolution of Cornubia. *Proceedings of the Ussher Society*, Vol. 5, 128–131.

— Harding, R R, and Darbyshire, D P F. 1975. Petrology and Rb: Sr age of the Brannel, South Crofty and Wherry elvan dykes, Cornwall. *Bulletin of the Geological Survey of Great Britain*, Vol. 52, 27–42.

Hendriks, E M L. 1931. The stratigraphy of south Cornwall. *Report of the British Association for 1930*, 332.

— 1937. Rock succession and structure in south Cornwall, a revision. With notes on the Central European facies and Variscan folding there present. *Quarterly Journal of the Geological Society, London*, Vol. 93, 322–360.

— 1939. The Start-Dodmon-Lizard Boundary Zone in relation to the Alpine structure of Cornwall. *Geological Magazine*, Vol. 76, 385–402.

— 1949. The Gramscatho Series. *Transactions of the Royal Geological Society of Cornwall*, Vol. 18, 50–61.

— 1959. A summary of present views on the structure of Cornwall and Devon. *Geological Magazine*, Vol. 96, 253–257.

— 1966. Correlation of south and north Cornwall. *Proceedings of the Ussher Society*, Vol. 1, 225–226.

— 1971. Facies variation in relation to tectonic evolution in Cornwall. *Transactions of the Royal Geological Society of Cornwall*, Vol. 20, 114–150.

— House, M R, and Rhodes, F H T. 1971. Evidence bearing on the stratigraphical successions in South Cornwall. *Proceedings of the Ussher Society*, Vol. 2, 270–275.

Henley, S. 1974. Geochemistry and petrogenesis of elvan dykes in the Perranporth area, Cornwall. *Proceedings of the Ussher Society*, Vol. 3, 136–145.

Henwood, W J. 1832. On some of the deposits of stream tin ore in Cornwall, with remarks on the theory of that formation. *Transactions of the Royal Geological Society of Cornwall*, Vol. 4, 58–69.

— 1843. On the metalliferous deposits of Cornwall and Devon. *Transactions of the Royal Geological Society of Cornwall*, Vol. 5. 512pp.

Hill, J B. 1913. The geology of northern Meneage. *Proceedings of the Geologists Association*, Vol. 24, 134–158.

— and MacAlister, D A. 1906. The geology of Falmouth and Camborne and of the mining district of Camborne and Redruth. *Memoir of the Geological Survey of Great Britain.*

Holder, M T, and Leveridge, B E. 1986a. A model for the tectonic evolution of south Cornwall. *Journal of the Geological Society, London*, Vol. 143, 125–134.

— — 1986b. Correlation of the Rhenohercynian Variscides. *Journal of the Geological Society, London*, Vol. 143, 141–147.

Holdsworth, R E, and Roberts, A M. 1984. Early curvilinear fold structures and strain in the Moine of the Glen Garry region, Inverness-shire. *Journal of the Geological Society, London*, Vol. 141, 327–338.

Holland, C H, and others. 1978. A guide to stratigraphical procedure. *Special Report of the Geological Society, London*, Vol. 10. 18pp.

Hosking, K F G. 1969. The nature of the primary tin ores of the south-west of England. 1157–1244 in *A second technical conference on tin.* Fox, W (editor). (London: International Tin Council.)

Institute Geological Sciences. 1975. Sheet 50°N, 06°W Land's End, Bouger gravity anomaly map. 1:250 000 series.

Jackson, N J, Halliday, A N, Sheppard, S M F, and Mitchell, J G. 1982. Hydrothermal activity in the St Just mining district. In *Metalization associated with acid magmatism.* Evans, A M (editor). (Chichester: Wiley.)

James, H C L. 1974. Problems of dating raised beaches in south Cornwall. *Transactions of the Royal Geological Society of Cornwall*, Vol. 20, 260–274.

— 1981a. Evidence for late Pleistocene environmental changes at Towan Beech, south Cornwall. *Proceedings of the Ussher Society*, Vol. 5, 238.

— 1981b. Pleistocene sections at Gerrans Bay, south Cornwall. *Proceedings of the Ussher Society*, Vol. 5, 239–240.

Jeffries, N L. 1985. The origin of sillimanite-bearing pelitic xenoliths within the Carnmenellis pluton, Cornwall. *Proceedings of the Ussher Society*, Vol. 6, 229–236.

Jenkyns, H C. 1978. Pelagic environments. 314–371 in *Sedimentary facies and environments.* Reading, H G (editor). (Oxford: Blackwell.)

Jones, P B. 1982. Oil and gas beneath east-dipping underthrust faults in the Alberta foothills. *Rocky Mountain Association of Geologists*, Vol. 6, 61–74.

Kidson, C. 1977. The coast of south west England. 257–298 *in* The Quaternary history of the Irish Sea. Kidson, C, and Tooley, M J (editors). *Geological Journal Special Issue*, No. 7.

LAMBERT, J L M. 1959. Cross-folding in the Gramscatho Beds at Helford River, Cornwall. *Geological Magazine*, Vol. 66, 489–496.

— 1965. A reinterpretation of the breccias in the Meneage crush zone of the Lizard boundary, south-west England. *Quarterly Journal of the Geological Society, London*, Vol. 121, 339–357.

LANG, W H. 1929. On fossil wood (*Dadoxylon hendriksi*, n.sp) and other plant remains from the clay-slates of S Cornwall. *Annals of Botany*, Vol. 43, 663–683.

LEAKE, R C, and others. 1985. Volcanogenic and exhalative mineralisation within Devonian rocks of the South Hams district of Devon. *Mineral Reconnaissance Report, British Geological Survey*, No. 79. 90pp.

LEDINGHAM, P. 1986. Heat mining. *Mining Magazine*, Vol. 154, 241–251.

LE GALL, B, LE HERISSE, A, and DEUNFF, J. 1985. New palynological data from the Gramscatho Group at the Lizard front (Cornwall): palaeogeographical and geodynamical implications. *Proceedings of the Geologists' Association, London*, Vol. 96, 237–253.

LEVERIDGE, B E. 1974. *The tectonics of the Roseland coastal section, south Cornwall.* Unpublished PhD thesis, University of Newcastle-upon-Tyne.

— 1987. *Geological description for 1:10 000 sheets SW84 NW, NE, SW, SE and parts of SW 94 NW and SW (Truro).* 51pp. (Keyworth: British Geological Survey.)

— and HOLDER, M T. 1985. Olistostromic breccias at the Mylor/Gramscatho boundary, south Cornwall. *Proceedings of the Ussher Society*, Vol. 6, 147–154.

— — and DAY, G A. 1984. Thrust nappe tectonics in the Devonian of south Cornwall and the western English channel. 103–112 in *Variscan tectonics of the North Atlantic region. Special Publication of the Geological Society, London*, Vol. 14, 103–112.

LISTER, C J. 1984. Xenolith assimilation in the granites of SW England. *Proceedings of the Ussher Society*, Vol. 6, 46–53.

MATTHEWS, S C. 1977. The Variscan fold belt in south-west England. *Neues Jahrbuch fur Geologie und Palaeontology*, Vol. 154, 94–127.

MCKEOWN, M C. 1966. Breccias of the Gorran Haven area. *Proceedings of the Ussher Society*, Vol. 1, 220–221.

MERRIMAN, R J. 1982. The mica crystallinity and mineralogical composition of slates from the Falmouth area, Cornwall. *Report of the Geochemistry and Petrology Division, Institute of Geological Sciences*, No. 219. 7pp.

MIDDLETON, G V, and HAMPTON, M A. 1976. Subaqueous sediment transport and deposition by sediment gravity flows. 197–218 in *Marine sediment transport and environmental management.* STANLEY, D J, and SWITT, D J P (editors). (New York: John Wiley.)

MITCHELL, A H G. 1974. Southwest England gravites: magmatism and tin mineralisation in a post-collision tectonic setting. *Transactions of the Institute of Mining and Metallurgy (Applied Earth Science)*, Vol. 83, 95–97.

MITCHELL, G F. 1973. The late Pliocene marine formation at St Erth, Cornwall. *Transactions of the Royal Society, London*, Vol. 266, 1–37.

MORRISON, T A. 1980. *Cornwall's central mines. The Northern District 1810–1895.* 392pp. (Penzance: Alison Hodge.)

MOTTERSHEAD, D N. 1977. The Quaternary evolution of the south coast of England. 299–320 *in* The Quaternary history of the Irish Sea. KIDSON, C, and TOOLEY, M J (editors). *Geological Journal Special Issue*, No. 7.

MURCHISON, R I. 1846. A brief review of the classification of the sedimentary rocks of Cornwall. *Transactions of the Royal Society of Cornwall*, Vol. 6, 317–326.

PANTIN, H M, and EVANS, C D R. 1984. The Quaternary history of the central and south western Celtic Sea. *Marine Geology*, Vol. 57, 259–293.

PEACH, C W. 1841. An account of the fossil organic remains found on the south-east coast of Cornwall, and in other parts of that country. *Transactions of the Royal Geological Society of Cornwall*, Vol. 6, (1864), 12–23.

PEARCE, J A, HARRIS, N B W, and TINDLE, A G. 1984. Trace element discrimination diagrams for the tectonic interpretation of granitic rocks. *Journal of Petrology*, Vol. 25, 956–983.

PHILLIPS, F C. 1964. Metamorphism in south-west England. 185–200 in *Present views of some aspects of the geology of Cornwall and Devon.* HOSKING, K F G, and SHRIMPTON, G J (editors). (Blackford, Truro: Royal Geological Society of Cornwall.)

QUINQUIS, H, AUDREN, C, BRUN, J P, and COBBOLD, P R. 1978. Intense progressive shear in the Ile de Groix blueschists and compatibility with subduction or obduction. *Nature, London*, Vol. 273, 43–45.

RAMSAY, J G. 1967. *Folding and fracturing of rocks.* 568pp. (New York: McGraw-Hill.)

— 1980. The crack-seal mechanism of rock deformation. *Nature, London*, Vol. 284, 135–139.

RATHORE, J S. 1980. A study of secondary fabrics in rocks from the Lizard peninsula and adjacent areas in southwest Cornwall, England. *Tectonophysics*, Vol. 68, 147–160.

RATTEY, R P. 1979. The relationship between deformation and intrusion of the Cornubian batholith in south west England. *Journal of the Camborne School of Mines*, Vol. 79, 60–63.

— 1980. *Structural studies in southwest Cornwall.* Unpublished PhD thesis, Queens University, Belfast.

— and SANDERSON, D J. 1982. Patterns of folding within nappes and thrust sheets: examples from the Variscan fold belt of SW England. *Tectonophysics*, Vol. 88, 247–267.

— — 1984. The structure of SW Cornwall and its bearing on the emplacement of the Lizard Complex. *Journal of the Geological Society, London*, Vol. 141, 87–95.

REID, C. 1890. The Pliocene deposits of Great Britain. *Memoir of the Geological Survey of Great Britain.*

— and FLETT, J S. 1907. Geology of the Land's End district. *Memoir of the Geological Survey of Great Britain.*

RHODES, S, and GAYER, R A. 1977. Non-cylindrical folds, linear structures in the X direction and mylonite developed during translation of the Caledonian Kala Nappe Complex of Finnmark. *Geological Magazine*, Vol. 114, 329–341.

RENOUF, J T. 1974. The Proterozoic and Palaeozoic development of the Armorican and Cornubian provinces. *Proceedings of the Ussher Society*, Vol. 3, 6–14.

RICKARD, M J. 1961. A note on cleavages in crenulated rocks. *Geological Magazine*, Vol. 98, 324–332.

RUPKE, N A. 1977. Growth of an ancient deep sea fan. *Journal of Geology*, Vol. 85, 725–744.

SADLER, P M. 1973. An interpretation of new stratigraphic evidence from South Cornwall. *Proceedings of the Ussher Society*, Vol. 3, 535–550.

Sanderson, D J. 1972. Oblique folding in south-west England. *Proceedings of the Ussher Society*, Vol. 2, 438–442.

Scrivener, R C. 1986. *Ore genesis in the Falmouth–Camborne district: a study of paragenesis and ore fluids at Wheal Jane Wheal Pendarves and South Crofty Tin Mine.* 25pp. (Keyworth, Nottingham: British Geological Survey.)

Sedgwick, A. 1852. On the slate rocks of Devon and Cornwall. *Quarterly Journal of the Geological Society, London*, Vol. 8, 1–19.

Selwood, E B, and Durrance, E M. 1982. The Devonian rocks. 15–41 in *The geology of Devon.* Durrance, E M, and Laming, D J C (editors). (Exeter: University of Exeter.)

Shackleton, R M, Ries, A C, and Coward, M P. 1982. An interpretation of the Variscan structures in SW England. *Journal of the Geological Society, London*, Vol. 139, 533–541.

Shaw, H R. 1965. Comments on viscosity, crystal settling, and convection in granitic magmas. *American Journal of Science*, Vol. 263, 120–152.

— 1972. Viscosities of magmatic silicate liquids: an empirical method of prediction. *American Journal of Science*, Vol. 272, 870–893.

Shepherd, T J, and Scrivener, R C. 1987. Role of basinal brines in the genesis of polymetallic vein deposits, Kit Hill —Gunnislake area, SW England. *Proceedings of the Ussher Society*, Vol.6, 491–97.

— Miller, M F, Scrivener, R C, and Darbyshire, D P F. 1985. Hydrothermal fluid evolution in relation to mineralization in southwest England with special reference to the Dartmoor–Bodmin area. 345–364 in *High heat production (HHP) granites, hydrothermal circulation and ore genesis.* (London: Institution of Mining and Metallurgy.)

Sheppard, S M F. 1977. The Cornubian batholith, SW England, D/H and 180/160 studies of kaolinite and other alteration minerals. *Journal of the Geological Society, London*, Vol. 133, 573–591.

Shipman, B. 1984. Healing the scars—the reclamation of derelict mine land in Cornwall. *Camborne School of Mines Journal*, Vol. 84, 59–62.

Smedley, P L, Bromley, A V, Shephard, T, and Edmunds, W M. In press. The hydrogeology and hydrogeochemistry of the Carnmenellis Granite, Cornwall. *British Geological Survey Technical Report.*

Smith, H G. 1929. Some features of Cornish lamprophyres. *Proceedings of the Geologists' Association*, Vol. 40, 260–268.

Smith, M A P. 1965. Repeated folding between Hayle and Portreath. *Proceedings of the Ussher Society*, Vol. 1, 170–171.

Snelling, N J (editor). 1985. The chronology of the geological record. *Memoir of the Geological Society, London*, Vol. 10.

Sorby, H C. 1858. On the microscopical structures of crystals indicating the origin of minerals and rocks. *Quarterly Journal of the Geological Society of London*, Vol. 14, 453–500.

South West Water Authority. 1979. Fortesque shaft, SW 668 389: hydrogeologist's report on pumping tests during 1979. Unpublished Report (Appendix III).

Stone, M. 1962. Vertical flattening in the Mylor Beds near Porthleven, Cornwall. *Proceedings of the Ussher Society*, Vol. 1, 25–27.

— 1966. Fold structures in the Mylor Beds, near Porthleven, Cornwall. *Geological Magazine*, Vol. 103, 440–459.

— 1968. A study of the Praa Sands elvan and ita bearing on the origin of elvans. *Proceedings of the Ussher Society*, Vol. 2, 37–42.

— 1979. Textures of some Cornish granites. *Proceedings of the Ussher Society*, Vol. 4, 370–379.

— and Austin, W G C. 1961. The metasomatic origin of the potash feldspar megacrysts in the granites of south-west England. *Journal of Geology*, Vol. 69, 464–472.

— and Exley, C S. 1978. A cluster analysis of chemical data from the granites of SW England. *Proceedings of the Ussher Society*, Vol. 4, 172–181.

Strong, D F, Stevens, R K, Malpas, J, and Badham, J P N. 1975. A new tale for the Lizard. (Abstract.) *Proceedings of the Ussher Society*, Vol. 3, 252.

Styles, M T, and Rundle, C C. 1964. The Rb-Sr isochron age of the Kennack Gneiss and its bering on the age of the Lizard Complex, Cornwall. *Journal of the Geological Society, London*, Vol. 141, 15–19.

Taylor, R T. 1963. An occurrence of cassiterite within a porphyry dyke at South Crofty Mine, Cornwall. *Transactions of the Institution of Mining and Metallurgy*, Vol. 72, 749–758.

— and Wilson, A C. 1975. Notes on some igneous rocks of west Cornwall. *Proceedings of the Ussher Society*, Vol. 3, 255–262.

Turner, R E, Taylor, R T, Goode, A J J, and Owens, B. 1979. Palynological evidence for the age of the Mylor Slates, Mount Wellington, Cornwall. *Proceedings of the Ussher Society*, Vol. 4, 274–283.

Turner, R G. 1968. The influence of granite emplacement on structures in south-west England. Unpublished PhD thesis, University of Newcastle-upon-Tyne.

Tyack, W. 1875. On a deposit of quartz gravel at Blue Pool, in Crowan. *Transactions of the Royal Geological Society of Cornwall*, Vol. 9, 177–181.

Walker, R G. 1978. Deep-water sandstone facies and ancient submarine fans: models for exploration for stratigraphic traps. *Bulletin of the American Association of Petroleum Geologists*, Vol. 62, 932–966.

Watson, J, Fowler, M B, Plant, J A, and Simpson, P R. 1984. Variscan–Caledonian comparisons: late orogenic granites. *Proceedings of the Ussher Society*, Vol. 6, 2–12.

Weller, M R. 1961. The palaeogeography of the 430 ft shoreline stage in East Cornwall [Abstract]. *Proceedings of the Fourth Conference of Geologists and Gemomorphologists in South West England*, 23–24.

White, S H, Burrows, S E, Carreras, J, Shaw, N D, and Humphreys, F J. 1980. On mylonites in ductile shear zones. *Journal of Structural Geology*, Vol. 2, 175–187.

Wintle, A G. 1981. Thermoluminescence dating of late Devensian loesses in southern England. *Nature, London*, Vol. 29, 479–480.

Wilson, A C, and Taylor, R T. 1976. Stratigraphy and sedimentation in West Cornwall. *Transactions of the Royal Geological Society of Cornwall*, Vol. 20, 246–259.

Ziegler, P A. 1982. *Geological Atlas of Western and Central Europe.* 130pp. (Moatschappij B V: Shell International Petroleum.)

APPENDIX 1

Selected boreholes

Abbreviated logs of six cored boreholes drilled in the district are given below. The National Grid reference and BGS registration number are given for each borehole.

BGS registration number **SW 63 NW/42** NGR 6363 3747

Surface level +83.82m Azimuth 150°N Inclination 45°

	Depth (m)
Overburden, including weathered slate	17.22
Mylor Slate Formation with intrusions	
Slate, spotted, fractured, altered down to 90.0	117.04
Elvan, pale to medium grey, phenocrysts of quartz and feldspar, chilled margin 0.86 m at top and 0.74 m at bottom	127.61
Slate, medium brown, lamined, spotted hornfelsed below 138.30 m	259.26
Granite, weathered, aplitic	262.20
Slate, medium brown, siltstone laminae, hornfelsed	268.45
Porthtowan Formation with intrusions	
Sandstone, medium brown, slaty in part, hornfelsed	290.45
Granite, fined-grained, aplitic	296.42
Sandstone, pale grey, hornsfelsed, intense below 329.28 m	349.58
Granite, coarse-grained, poorly megacrystic, biotitic, megacrysts up to 20 mm more abundant below 350.19 m	357.53
Slate, greyish brown, laminated, hornfelsed	359.18
Granite, xenoliths of hornfels	360.63
Sandstone, slate interbeds, grey to brown with many thin granite veins, hornsfelsed	400.99
Granite, megacrystic, biotitic, aplitic between 411.05 and 417.42 m	420.60
Greenstone, fine-grained	466.65
Slate, with siltstone and sandstone laminae, medium to dark grey, hornfelsed	484.10
Greenstone, medium- to coarse-grained, hornfelsed	487.53
Slate, medium greyish brown, with siltstone laminae banded, hornfelsed	490.88
Granite, medium-grained, partly kaolinitised, 0.23 m chilled margin, variously altered	552.98

BGS registration number **SW 63 NW/51** NGR 6157 3633

Surface level +82.30 m

	Depth m
Overburden	0.91
Head	4.65
Mylor Slate Formation with intrusions	
Slate, dark grey, smooth, weathered to 11.28 m, vaguely spotted below 23.95 m	50.43
Intrusive breccia, fine-grained, kaolinised, coarser at base	50.68
Slate, darkish grey, smooth	52.00
Intrusive breccia, granitic matrix, some rounded fragments of slate	52.33
Dark slate, as above	52.59
Intrusive breccia, as above	52.69
Slate, dark grey, smooth, locally well spotted below 67.34 m, some siltstone beds, particularly below 140.00 m. Sandstone at 157.53 m to 157.86 m	233.86
Porthtowan Formation with intrusions	
Slate, dark grey to brownish grey, interbedded with paler grey siltstone and sandstone	260.63
Elvan, fine-grained, equigranular to finely porphyritic	297.92
Intrusive breccia, angular to rounded clasts of various silty rocks, vein quartz and rare granite in a granitic clastic matrix	299.77
Elvan as above	305.66
Intrusive breccia as above	305.66
Elvan, fine-grained, grey with glassy groundmass	308.15
Sandstones, pale grey, bedded with darker grey slate	311.20
Elvan, fine-grained, pale grey groundmass at margin, within 75 mm becoming coarsely porphyritic, feldspars white and fresh up to 15 mm	311.96
Sandstone as above	312.29
Elvan, fine-grained, porphyritic feldspars up to 15 mm	315.39
Greenstone, fine-grained	320.68
Elvan, as above	329.54
Greenstone, fine-grained	332.87
Elvan, as above	337.52
Greenstone, fine-grained	339.62
Sandstone, pale grey, interbedded with dark greenish or brownish grey slate, spotted in part, hornfelsed, intruded by greenstone at 343.08 m to 343.38 m, 355.93 m to 359.00 m, 383.90 m to 385.22 m, 383.75 m to 386.51 m, 395.48 m to 395.73 m, 396.09 m to 396.16 m, 410.54 m to 414.32 m, 434.16 m to 434.31 m	513.59
Granite, fine-grained, aplitic, greisened	514.17
Sandstone, pale grey, interbedded with dark lustrous or micaceous, grey to brown, hornfelsed slate, graded bedding becomes common	584.02
Granite, fine-grained, aplitic, pegmatitic at top	584.40
Sandstone, pale grey with thin darker grey hornfelsed slate laminae	587.10
Greenstone, fine-grained	594.39
Granite, fine-grained	595.25
Slate, adinolised	595.33
Greenstone, fine-grained	596.01
Slate, biotitic, hornfelsed	596.19
Greenstone, fine-grained	597.18
Sandstone, pale grey with interbedded biotitic slates, hornfelsed	597.59
Greenstone, fine-grained	597.97
Sandstone, pale grey, turbiditic, interbedded with thinner greenish brownish, or dark grey hornfelsed slate	665.51

BGS registration number **63 NW/82** NGR 65849 36609

Surface level + 155.0 m Inclination 90°

	Depth m
NO CORE RECOVERED	73.00
G_a granite, coarse-grained megacrystic with feldspar megacrysts up to 20 mm long in planar alignment. Matrix of quartz, feldsapr biotite ± white mica. In part the feldspar megacrysts have been replaced by white mica in zones, fractures and at intergrain boundaries. Rounded micaceous xenolith at 81.00 m	389.00
Elvan, quartz-feldspar-porphyry ± biotite, fine-grained chilled margin becoming megacrystic over 1.0 m. Feldspar megacrysts bladed up to 15 mm long. Angular biotite-rich schistose xenoliths up to 50 mm long and sub-angular fine- to medium-grained nonmegacrystic granite xenoliths less than 50 mm long present. Irregular margins and iron-stained centre to elvan	400.00
G_a granite as above	
Elvan, as above with biotite in matrix, generally iron-stained and with possible sedimentary rock xenoliths	502.50
G_d granite with fine- to medium-grained matrix and quartz and feldspar as above. Tourmaline breccia veins at base 0.5 m thick	546.50
G_a granite as above 389.90 m but grading down into G_d phase below 550 m and back into G_a phase at 591.4 m with veins of fine-grained granite	600.00
G_d granite, megacrystic granite with rather variable fine- to medium-grained matrix	

BGS registration number **SW 64 SE/83** NGR 6742 4187

Surface level + 105 m

	Thickness m	*Depth* m
MYLOR SLATE FORMATION with intrusions		
Slate, weathered and decomposed at top passing down into grey slate with dark spots composed of biotite, tourmaline and quartz. Films of mica along cleavage planes	67.00	67.00
Hornfels, foliated, biotitic with a zinc blende-galena-pyrite lode at top, dipping north at the same angle as the slate and hornfels. Films of Zn, Pb, Fe sulphides on cleavage planes ?45°. Becomes harder and darker towards base	36.63	103.63
Greenstone, fine-grained, altered, dipping ?20° to north, veined with brown garnet and pale green pyroxene, patches of axinite, pyrrhotite and chalcopyrite. Lower part penetrated by narrow veins of elvan	57.91	161.54
Elvan, buff-coloured, megacrystic, dipping north at ?55–60°. Megacrysts of bipyramidal quartz, pink orthoclase feldspar and patches of radiating tourmaline needles. Microcrystalline groundmass, granite vein, (depth not known) cuts elvan and consists of two parts: an upper coarsely crystalline pegmatite with crystals showing vague outlines and a well-developed graphic texture. Quartz muscovite, orthoclase, lepidolite and chlorite with fine needles of tourmaline. A lower part consists of fine-grained granite, similar in composition to above but with some biotite, sodium and sodium-calcium feldspar. Immediately below the elvan a lode containing pyrite, chalcopyrite, blende arsenopyrite with ankerite, fluorite chlorite and tourmaline was drilled	27.44	188.98
Greenstone, as above with some dark grey-black spotted slate horizons near top, passing into continuous greenstone below 231.70. Hornfelsed slate between 261.52–268.83 m	137.16	326.14
Hornfelsed slate as above	26.82	352.96
Granite to base of shaft	c.220.06	c.573.02

BGS registration number **SW 64 SE/84** NGR 6860 4231
(New Tolgus Shaft)

Surface level + 104 m

	Thickness m	*Depth* m
SOIL AND HEAD	9.1	9.1
MYLOR SLATE FORMATION		
Slate, blue-grey, decomposed with indistinct spotting which becomes stronger below 36.58 m. Below 61 m slate becomes paler blue-grey with green beds	71.1	80.80
Elvan, dipping north at c.40°, much decomposed	7.32	88.12
Slate, dark blue-grey, green in part, spotted. Below 122.0 m becomes yellowish grey clayey slate	55.14	143.26
PORTHTOWAN FORMATION?		
Greywacke and grit, greenish, foliated calcareous and pyritiferous	3.04	146.30
Slate, blue-grey, spotted	14.63	160.93
Intrusive breccia with angular clasts of slate and elvan from 0.02 to 0.31 m. Elvan resembles that between 80.80 and 88.12 m, in this shaft. Feldspathic and calcareous matrix	12.81	173.74
Elvan, fine-grained, micrograninic dipping 43° north. White in colour with small sporadic crystals of chlorite and tourmaline. Upper chilled margin c.0.31 m thick	22.86	196.60
Slate, with quartz bands and chlorite	3.04	199.64
Elvan, pink megacrysts of feldspar, dips north at c.40°. Soft and much decomposed	6.1	205.74
Intrusive breccia described as 'conglomerate' and consists of rounded clasts of clay slate, biotite-hornfels, quartz and elvan	7.62	213.36
Greenstone, fine-grained with much axinite, garnet, zoisite and epidote. Top 10 m highly mineralised	84.73	298.09
Slate, dark brown to black, biotitic, hornfelsed. A tongue of schorl-rock and greisen in South side of shaft at 307.85 m. Black, spotted clay slate below. At 329.18 m a vein of brecciated dolerite with scattered arsenopyrite crystals. Continuing below in lustrous phyllite with increasing amounts of white mica	157.59	455.68

Pegmatitic granite vein with white and pink feldspar crystals, quartz, muscovite, tourmaline and apatite	0.38	456.06
Aplitic granite, veined with cassiterite	4.49	460.55
Phyllite as above, mineralised between 463.30 and 481.58 m	13.41	473.96
Slate, dark grey, hornfelsed, interbedded with greenstone. At 489.20 m a tongue of graphic granite was seen on one side of shaft. At 531.8 m a narrow south-dipping aplite was cut	53.35	537.06
Greenstone	3.05	540.11
Aplite	3.04	543.15
Slate, hornfelsed, and greenstone interbedded. At 593.75 m an aplite vein 6.1 m thick dipping 50° north	65.23	608.38
Elvan, north dipping to bottom of shaft	1.22	609.60

BGS registration number **SW 74 SW/11** NGR 7024 4374

Surface level + 106.7 m Azimuth 337° Inclination 43°

	Thickness m	*Depth* m
PORTHTOWAN FORMATION		
Sandstone, interbedded with thin slate and siltstone	11.89	14.63
Sandstone or grit, coarse-grained, orange and red stained, cleaved at 60–70° to core length	4.73	19.36
Sandstone, siltstone and slate, dark grey, becoming less arenaceous, some graded beds are inverted	14.17	33.53
Slate, grey, spotted, and siltstone	9.6	43.13
Sandstone	2.59	45.72
Slate and siltstone	15.39	61.11
Sandstone, fine-grained, cleaved with inclusions of quartz-veined slate	3.2	64.31
Slate, grey black, well-cleaved, quartz veined and spotted	1.27	65.58
Sandstone, fine-grained, grey	5.13	70.71
Slate and siltstone	30.33	101.04
Sandstone, interbedded with slate	5.64	106.68
Slate, grey, spotted and siltstone with minor sandstone	182.88	289.56

APPENDIX 2

BGS publications and data relevant to the area

1 Maps at 1:10 000 scale, solid editions

Manuscript copies of all the sheets are available for consultation only, but dyeline copies of some sheets may be purchased. These maps are derived from surveys between 1974–84.

Maps available as manuscript and dyeline copies

SW 63 NW
SW 63 NE
SW 63 SW
SW 63 SE
SW 63 NE
SW 64 SW and part of SW 64 NW
SW 64 SE
SW 73 NW
SW 73 NE
SW 73 SW
SW 73 SE
SW 74 NW
SW 74 NE
SW 74 SW
SW 74 SE
SW 83 NW
SW 83 NE
SW 83 SW
SW 83 SE
SW 84 NW
SW 84 NE
SW 84 SW
SW 84 SE

Maps available as manuscript copies only

Part of SW 62 NW
Part of SW 62 NE
Part of SW 72 NW
Part of SW 72 NE and part of SW 82 NW
Part of SW 93 NW
Part of SW 94 NW
Part of SW 94 SW

2 Maps at 1:50 000 scale

Falmouth (Sheet 352) Solid and Drift edition 1989

3 Maps at 1:253 440 (¼-inch-to-one-mile) scale

Bodmin, Truro, Falmouth, Land's End, Isles of Scilly (Sheet 21 and 25) Solid edition 1969.

4 Maps at 1:250 000 scale

Land's End (Sheet 50°N-06°W) Solid edition 1984
Land's End (Sheet 50°N-06°W) Aeromagnetic Anomaly Map, 1977
Land's End (Sheet 50°N-06°W) Bouguer Gravity Anomaly Map, 1975

5 Memoirs

The Metalliferous Mining Region of South-West England by H G Dines, with notes by J Phemister. 2 Volumes. First published 1956. Second impression with amendments 1969. (Out of print.)
The geology of Falmouth and Camborne and of the Mining district of Camborne and Redruth. J B Hill and D A MacAlister, with petrological notes by J S Flett. 1906
Geology of the country around Penzance. A J J Goode and R T Taylor. 1988

6 British Regional Geology

South-West England by E A Edmonds, M C McKeown and M Williams. (fourth edition) 1975.

7 IGS Report Series

73/7 The mode of intrusion of Cornish elvans. A J J Goode, 1973.

8 Open File Reports

Geological notes and details for 1:10 000 sheets SW 62 NE and NW (part) and sheets SW 63 NW, NE, SW and SE (Crowan, Cornwall). M T Holder, 1987.
Geological notes and details for 1:10 000 sheets SW 64 NW, NE, SW and SE (Camborne–Redruth, Cornwall). A J J Goode, 1987.
Geological notes and details for 1:10 000 sheets SW 72 NE and NW (part) and sheets SW 73 NW, NE, SW and SE (Stithians, Cornwall). M T Holder, 1987.
Geological notes and details for 1:10 000 sheets SW 74 NW, NE, SW and SE (Chacewater, Cornwall). A J J Goode, 1987.
Geological notes and details for 1:10 000 sheets SW 83 NW, NE, SW and SE and parts of SW 93 NW and SW (Falmouth and south Roseland, Cornwall). B E Leveridge, 1987.
Geological notes and details for 1:10 000 sheets SW 84 NW, NE, SW and SE and parts of SW 94 NW and SW (Truro, Cornwall). B E Leveridge, 1987.
Ore genesis in the Falmouth–Camborne district: a study of paragenesis and ore fluids at Wheal Jane, Wheal Pendarves and South Crofty Tin Mine. R C Scrivener, 1987.

Geothermal Energy Research Programme

The origin and circulation of groundwater in the Carnmenellis granite: the hydrogeochemical evidence. W G Burgess, W M Edmunds, J N Andrews, R F L Kay and D J Lee, 1982. 87pp.

Analytical Chemistry Research Group

No. 39 Chemical analysis of core samples from tin-bearing killas from Wheal Jane Mine, Cornwall. N Cogger, R A Nicholson, W G Prewett, L M Rundle and E Waine, 1969. 3pp.
No. 46 Chemical analysis of core samples of tin-bearing killas

from Wheal Jane Mine, Cornwall. Part 2. N Cogger, R A Nicholson and E Waine, 1970. 3pp.
No. 53 The analysis of mine waters from Cornwall. R A Nicholson and L M Rundle, 1970. 2pp.

Geochronology and Isotope Geochemistry Research Group

No.70.27 The Rb-Sr and the K-Rb ratio of an elvan dyke in the South Crofty mine, Pool, Cornwall. R R Harding and J R Hawkes, 1970. 5pp.
No.77.13 K:Ar age determinations on two lamprophyre dykes from S W England. C C Rundle, 1973. 2pp.
No.80.7 K-Ar from lamprophyre dykes from S W England. C C Rundle, 1980. 3pp.

Mineral Resources and Applied Geochemistry Research Group

No. 94 Report on a radiometric reconnaissance of old mine dumps in Cornwall and Devon. D Ostle, 1950. 7pp.
No. 213 The occurrence of uranium at Roskrow United Mine, Ponsanooth, Cornwall. J Taylor, 1959.

Mineral Reconnaissance Programme Report Series

No. 1 The concealed granite roof in south-west Cornwall. K E Beer, A J Burley and J M Tombs, 1975. 15pp.
No. 2 Geochemical and geophysical investigations around Garras Mine, near Truro, Cornwall. R C Jones and J M C Tombs, 1975. 14pp.
No. 11 A study of the space form of the Cornubian granite batholith and its application to detailed gravity surveys in Cornwall. J M C Tombs, 1977. 16pp.

Mineralogy and Petrology Research Group

No. 58 Particle size analyses of mud fractions of 26 ?Pleistocene samples from Pendarves, Cornwall. G E Strong, 1973. 3pp.
No. 219 The mica crystallinity and mineralogy composition of slates from the Falmouth area, Cornwall. R J Merriman, 1972. 7pp.

Engineering Geology and Reservoir Rock Properties Research Group

No.EG79/5 Acoustic studies in the Carnmenellis granite at Rosemanowas Quarry, Cornwall. D M McCann and R Baria, 1979. 5pp.

Global Seismology Research Group

No.189 Monitoring the background and induced seismicity for the Hot Dry Rock programme in Cornwall. T Turbitt, A B Walker, C W A Browitt and S N Morgan, 1973.

9 Photographs

Albums of coloured and black and white prints, illustrating the geology of the district, are available for consultation in BGS libraries. All photographs are also available as 35 mm colour slides.

10 Rock samples

Rock samples and microscope sections used in the description of the geology of the district are available for examination on request.

11 Boreholes

Deep boreholes, drilled for the purposes of locating mineral veins, and the results of some limited shallow drilling are available for consultation in BGS offices in Exeter or Keyworth.

12 Consultation

Enquiries regarding the geology of the district should be directed to:

British Geological Survey
St Just,
30 Pennsylvania Road,
Exeter EX4 6BX
Tel. (0392) 78312

Material may be inspected, by appointment only, at the Exeter Office or via the National Geosciences Database at:

British Geological Survey,
Keyworth,
Nottingham NG12 5GG
Tel. (06077) 6111

APPENDIX 3

Viscosity and magma flow velocities for the Carnmenellis granite calculated from the method of Shaw (1972).

Estimates of the flow velocity within the Carnmenellis Granite can be made by assuming that country rock xenoliths within it were in free fall prior to freezing in during magma consolidation, and that the largest xenoliths present at outcrop are therefore the largest that could be supported by an upwelling magma current. This assumes that the xenoliths were not captured at random as the freezing interface moved inwards during cooling. Random capture of the xenoliths should lead to the presence of a wide size range of xenoliths distributed within the granite. As the observed size range of xenoliths in the Carnmenellis Granite is 10–100 mm some degree of sorting appears to have taken place implying the presence of some current activity.

Stokes law states that the velocity (v) of a spherical body falling in a liquid is given by:

$$v = \frac{2gr^2 (Dp)}{9n}$$

where
r is the radius of the body
Dp is the density difference between liquid and body
n is the viscosity of liquid
g is the gravity acceleration

This holds only if (Dp) lies between 0.1 & 1.0 g/cc and the Reynolds number (Re) is less than 0.05 (Shaw 1965).

$$Re = \frac{2pr}{n}$$

where
p is the density of body
r is the radius of body
n is the viscosity of liquid

Assuming a density for the granite magma of 2.3 g/cc the maximum value of r for Re = 0.05 is 3200 m. This is much greater than the maximum size of xenolith present in the pluton (100 mm). Assuming Dp is 0.15 g/cc (Bott and others, 1970) the sinking velocity of the xenoliths can be determined. In order for a xenolith to remain near the top of the granite its rate of fall through the magma must be less than, or equal to the upward flow of the magma. The rate of fall of the largest xenolith at the present level of outcrop must, therefore, be equal to the magma flow velocity, larger xenolith having fallen to the bottom of the intrusion. The rate of magma flow in the Carnmenellis Granite is therefore equal to the settling velocity of a 100mm diameter xenolith.

Water content of average granite %	Flow velocity
at 750°C	
0	3.4×10^{-11} m/s
1	6.8×10^{-10} m/s
2	1.3×10^{-8} m/s
3	1.0×10^{-7} m/s
4	6.3×10^{-7} m/s
5	3.7×10^{-6} m/s
at 1000°C	
0	6.8×10^{-8} m/s
1	7.4×10^{-7} m/s
2	6.8×10^{-6} m/s
3	3.4×10^{-5} m/s
4	1.4×10^{-4} m/s
5	5.4×10^{-4} m/s

INDEX

Page numbers in italics refer to illustrations and tables

BRITISH GEOLOGICAL SURVEY

Keyworth, Nottingham NG12 5GG
(06077) 6111

Murchison House, West Mains Road,
Edinburgh EH9 3LA   031-667 1000

London Information Office, Natural History Museum,
Earth Galleries, Exhibition Road, London SW7 2DE
071-589 4090

The full range of Survey publications is available through the Sales Desks at Keyworth, Murchison House, Edinburgh, and at the BGS London Information Office in the Natural History Museum, Earth Galleries. The adjacent bookshop stocks the more popular books for sale over the counter. Most BGS books and reports are listed in HMSO's Sectional List 45, and can be bought from HMSO and through HMSO agents and retailers. Maps are listed in the BGS Map Catalogue and the Ordnance Survey's Trade Catalogue, and can be bought from Ordnance Survey agents as well as from BGS.

The British Geological Survey carries out the geological survey of Great Britain and Northern Ireland (the latter as an agency service for the government of Northern Ireland), and of the surrounding continental shelf, as well as its basic research projects. It also undertakes programmes of British technical aid in geology in developing countries as arranged by the Overseas Development Administration.

The British Geological Survey is a component body of the Natural Environment Research Council.

Maps and diagrams in this book use topography based on Ordnance Survey mapping

HMSO publications are available from:

HMSO Publications Centre
(Mail and telephone orders)
PO Box 276, London SW8 5DT
Telephone orders 071-873 9090
General enquiries 071-873 0011
Queueing system in operation for both numbers

HMSO Bookshops
49 High Holborn, London WC1V 6HB
071-873 0011 (Counter service only)
258 Broad Street, Birmingham B1 2HE
021-643 3740
Southey House, 33 Wine Street, Bristol BS1 2BQ
(0272) 264306
9 Princess Street, Manchester M60 8AS
061-834 7201
80 Chichester Street, Belfast BT1 4JY
(0232) 238451
71 Lothian Road, Edinburgh EH3 9AZ
031-228 4181

HMSO's Accredited Agents
(see Yellow Pages)

And through good booksellers